THE AUTHORITY GAP

www.penguin.co.uk

THE
AUTHORITY GAP

Why women are *still* taken less seriously
than men, and what we can do about it

Mary Ann Sieghart

doubleday

TRANSWORLD PUBLISHERS
Penguin Random House, One Embassy Gardens,
8 Viaduct Gardens, London SW11 7BW
www.penguin.co.uk

Transworld is part of the Penguin Random House group of companies
whose addresses can be found at global.penguinrandomhouse.com

Penguin
Random House
UK

First published in Great Britain in 2021 by Doubleday
an imprint of Transworld Publishers

A CIP catalogue record for this book
is available from the British Library.

ISBNs 9780857527561 (cased)
9780857527578 (tpb)

Typeset in 12.25/16 pt Dante MT Pro by Jouve (UK), Milton Keynes
Printed and bound in Great Britain by Clays Ltd, Elcograf S.p.A.

The authorized representative in the EEA is Penguin Random House Ireland,
Morrison Chambers, 32 Nassau Street, Dublin D02 YH68.

Penguin Random House is committed to a sustainable
future for our business, our readers and our planet. This book
is made from Forest Stewardship Council® certified paper.

FSC
www.fsc.org

MIX
Paper from
responsible sources
FSC® C018179

To Dai, the unlikely feminist

Contents

Introduction
Why Bart Simpson has more authority than Margaret Thatcher

THE AUTHORITY GAP? AH YES. ALLOW ME TO EXPLAIN. IT MEANS...

He tells her that the earth is flat –
He knows the facts, and that is that.
In altercations fierce and long
She tries her best to prove him wrong,
But he has learned to argue well.
He calls her arguments unsound
And often asks her not to yell.
She cannot win. He stands his ground.
The planet goes on being round.

– Wendy Cope

WHEN MARY MCALEESE was President of the Republic of Ireland, she led an official visit to the Vatican to meet Pope John Paul II. She was in the audience room at the head of her delegation, about to be introduced to the pontiff, when he reached straight past her, held out his hand to her husband instead, and asked him, 'Would you not prefer to be President of Ireland rather than married to the President of Ireland?' Her husband knew better than to take the bait. As McAleese told me in an interview for this book: 'I reached and took the hand which was hovering in mid-air and said, "Let me introduce myself. I am the President of Ireland, Mary McAleese, elected by the people of Ireland, whether you like it or whether you don't."'[1]

The Pope later claimed it was a joke but, if so, it was in poor taste. He had managed to snub a head of state before even acknowledging her presence. As McAleese recalled: 'He said, "I'm sorry, I tried a joke because I heard you had a great sense of humour." I said, "I do, but that wasn't funny because you would not have done that to a male President."' We would automatically respect a male head of state, so why not a female one? Because there is still an *authority gap* between women and men.

This sort of behaviour is incredibly frustrating for women. No one likes to be treated as if they're inferior, particularly if they're not. To see this more clearly, it helps to flip things around. So, if you're a man, I would like you to take a moment or two to do this thought experiment. Imagine living in a world in which you were routinely patronized by women. Imagine having your views ignored or your expertise frequently challenged by them. Imagine trying to speak up in a meeting, only to be talked over by female colleagues. Imagine women subordinates resisting you as a boss, merely because of your gender. Imagine women superiors promoting other women, even if they are less talented than you. Imagine people always addressing the woman you are with before they address you. Imagine writing a book and finding that half the population is reluctant to read it because it is written by a man. Would you just shrug your shoulders and say, 'Well, that's fair enough. Men are different from women.' Or would it infuriate you? I think I can guess.

R.E.S.P.E.C.T is what the soul singer Aretha Franklin demanded, and it's what women still have to fight harder than men to earn. However much we claim to believe in equality, we are still, in practice, more reluctant to accord authority to women than to men, even when they are leaders or experts. Every woman has a tale to tell about being underestimated, talked over, ignored, patronized and generally not taken as seriously as a man. (And when I say 'woman', I mean anyone who identifies as a woman.) Great strides have been made, and many men are good and respectful listeners, but however liberal we think we are, we're nowhere near there.

Research shows that we still expect women to be less expert than men. Most of us – men and women – are still less willing to be influenced by women's views. And we still resist the idea of women having authority over us. In other words, there is still an authority gap between women and men.

And the authority gap is the mother of all gender gaps. If women aren't taken as seriously as men, they are going to be paid less, promoted less and held back in their careers. They are going to feel less confident and less entitled to success. If we don't do anything about it, the gap between women and men in the public sphere will never disappear.

That gap is both huge and unmerited. The difference between

the amount that men and women are paid and promoted is fourteen times greater than the difference between their performance evaluations. This is because 70 per cent of men rate men more highly than women for achieving the same goals.[2] And in really prestigious jobs, in professions and senior management, women and men perform equally well, but women are paid significantly less.[3]

This is, at least in part, because we are still too ready to associate 'male' with 'authority'. When I put together some slides to give a lecture on this subject at Oxford University, I knew I had to start by defining my terms. So I took a screenshot of the Oxford Dictionary Online's definition of 'authority' – the very first result to come up on my Google search.[4] Every sentence it offered as an example began with the same pronoun: '*He* had absolute authority over his subordinates'; '*He* has the natural authority of one who is used to being obeyed'; '*He* hit the ball with authority'; and '*He* was an authority on the stock market.' I couldn't have found a neater illustration of the problem. Yet didn't Margaret Thatcher have the natural authority of one who was used to being obeyed? Doesn't Serena Williams hit a ball with authority? Isn't Helena Morrissey an authority on the stock market? I wasn't looking for this differential treatment, only for a dictionary definition.

The same happened when I searched in Google Images for something to illustrate a slide on 'expertise'. In the first twenty pictures, there wasn't a single woman. Bart Simpson appeared before we reached the first female, in a group with men. Finally, there was a decent-sized photo of a woman, but it turned out that she was having something explained to her by a male expert. Sometimes your subject just jumps up and slaps you in the face.

Surely this is all changing now, though? We're at last appointing more women to top jobs, and hurray for that. We now berate the Academy Award voters for ignoring female directors and are thrilled when more meaty parts are written for women. But what we've been seeing in the developed world since #MeToo is a kind of lip-service feminism. We are still more likely to follow and retweet men than women on Twitter. We are still more likely – when we walk up to a man and woman standing together – to address the man first. Men are still more inclined to ignore books written by women, though women lap up books written by men.

Unconscious bias seems to dog each step forward we take, and we're far too ready to congratulate ourselves for the progress we've made and ignore, or fail to notice, the bias that still exists. In this book, I want to examine our biases in detail and map out the measures we can take, as individuals and as a society, to spot them, counteract them, and see them for what they are: an irrational and anachronistic product of social conditioning and outdated stereotypes.) ' in her book, she aims to '_"...'

Our brains are used to taking short cuts – what psychologists call 'heuristics' – and dividing the world into categories so that we don't have to process too much information. We overlay these templates on to something like a transparent film between us and the person we're interacting with. Instead of treating each person as an individual, we map on to them our assumptions about what they *should* be like or what we *expect* them to be like, based on the stereotypes we've been brought up with and are surrounded by. We associate men with leadership in the outside world and women with home. These stereotypes may bear no resemblance to the actual person in front of us, but that doesn't stop us applying them. As Helle Thorning-Schmidt, the former Danish prime minister, put it to me: 'We are people who walk around with a brain that is wired to be extremely prejudiced against female leaders because it just goes against the grain of what our Stone Age brain can capture.'[5] It doesn't have to be evolutionarily determined from the Stone Age, though. It can be socially constructed from the contemporary world.

It's easy to underestimate how fundamental these templates are. We may have a sincere moral and intellectual objection to women being treated any differently from men. But the trouble is that once we've learned to see the world in a particular way, we may not even be aware of our subliminal prejudices. They are conveyed to us at too fundamental a level – as scientists of consciousness would explain, they are inextricably woven into the way we perceive the world.

Of course, there are also authority chasms between white and minority ethnic people, between people from different classes, between the able-bodied and the disabled, and between straight people and those with other sexualities. Each of these gaps deserves a whole book of its own, but I'm not the right person to write those books. So I'm mainly confining myself here to the gender authority gap, though I do also explore how it intersects with other biases.

The authority gap affects women all over the world, whatever the differences in culture. I've talked to women from Africa, Latin America, Asia and the Middle East, as well as from Europe and America, and they all say that they have experience of being taken less seriously than men. We might not notice that we interrupt them more, challenge their expertise more and listen to them less. But many of us do, and it's both insulting and wrong. It makes women fume, it dents their confidence and it holds them back. It's high time we changed our behaviour.

Even the most senior of women experience it, particularly women of colour. I talked to Bernardine Evaristo, Booker Prize-winning novelist and Professor of Creative Writing. Despite her impeccable credentials, her evident brilliance and her charismatic personality, she still has to fight to persuade people to accord her the authority she deserves, especially a certain type of older white male student 'who has come to learn from me but doesn't really believe that I have anything to teach him and will then challenge everything I say'.

'I have positions in my life where I have authority,' she told me. 'I am a professor at a university, and I am vice-chair of the Royal Society of Literature. I have now won the Booker. I have reviewed books and written essays for newspapers for decades. So on the outside you think, "Well, this is somebody who walks with authority," but actually, that's not how I'm treated by society. I'm very aware of it, that people don't automatically give me the authority that they would give someone else who they think should be in the positions that I hold.'[6]

And it's not just men who do it. This is not a man-bashing book. It's a consciousness-raising exercise for us all. For however progressive and intelligent we think we are, innumerable scientific studies show that we all – women as well as men – have unconscious biases, even against our own gender.[7] We may not be aware of them – they are called 'unconscious' for a reason – but they spill out into our behaviour and, unless we notice and correct for them, we will continue to take women less seriously than men. We will continue to assume that a man knows what he's talking about until he proves otherwise, while for a woman it's all too often the other way around. The authority gap will remain as wide as ever.

I'm not exempt myself – I'm not sure that any of us are. I'm a lifelong feminist and I've written a book begging the world to take women more seriously, but I too can suffer from this bias. Sometimes I'll hear a young woman being interviewed on the radio, who maybe has a high voice and sounds a bit childish in a way that men never do because their voices break, and I'll find myself thinking, 'I wonder if she knows what she's talking about?'

Of course, I immediately feel guilty and try to compensate. I'll listen carefully to what she's saying and then judge her on the content of her speech, not the pitch of her voice. Because it's only if we spot our bias in the first place, and then actively correct for it, that things can begin to change.

Over the course of writing this book, I've interviewed about fifty of the world's most powerful, successful – and authoritative – women to see whether even they have experience of the authority gap. Frances Morris is right at the top of the art world. She is the Director of Tate Modern. But that only insulates her from the authority gap when people already know what she does. We met in her office, brimming with books, at the vast London gallery. She is startlingly intelligent and talks in fully formed paragraphs. 'As director of Tate Modern, I can spend all day as a powerful, articulate person who's taken seriously, and I can leave this building and I am nobody,' she told me.[8] 'Because as a woman in the world out there, I'm not taken seriously. I'm very often in situations where people don't know my job and my hand is shaken after the hand of a male colleague, my eyes are met after the eyes of my husband are met, and no interest is shown in my opinion if it's just my opinion as a late middle-aged female. I see all that because it's in stark contrast with the way I'm approached and treated when it's known I'm the Director of Tate Modern.'

Might that not also be true of men in a similar position? 'No,' she insisted. 'I've worked for three male directors of Tate Modern, and it hasn't happened to them.'

I'm sure that Frances Morris is an avowed feminist. I doubt that Elaine Chao is. But you don't have to be a feminist to notice or care about these things. When I met her, Chao was US Transportation Secretary and a member of President Trump's Cabinet. We talked in her Washington office, with two of her long-standing senior female aides. Exuding confidence, and clearly one of those women who is used to dominating a room, she told me it had taken her until her late forties before she felt she was being listened to as much as her male colleagues. It finally happened when she became Labor Secretary under George W. Bush. 'Isn't that amazing?' she asked. 'I was forty-seven years old. For the first time in my life, I felt as if I had kind of made it.'[9] But she had held several top leadership jobs before then, including Director of the US Peace Corps, so if she wasn't listened to properly in those positions, it is a sign of how pervasive the authority gap is.

Women, even very senior ones, are so used to manifestations of the authority gap that they are pleasantly surprised when it doesn't show. Baroness Hale, then President of the UK Supreme Court, told me that she was taken aback when she and her (male) deputy hosted a meeting and the visitor, for once, addressed most of his questions to her rather than her deputy. It made her realize how unusual this was, even though she was the most senior person in the room – and in the whole judiciary.[10]

So why tell these top women's stories? If even super-powerful women have had their authority challenged, their views ignored and their expertise questioned, then it's a pretty good indicator that the rest of womankind have too. If they have managed to overcome this problem, their experiences can be useful to the rest of us, with much less privilege, from all backgrounds, in all walks of life.

But I've also talked to less well-known women, a diverse range across age, race and class. In the process, I haven't met a single one who hasn't encountered this phenomenon, from baby boomers through to Gen Xs, millennials and Gen Zs. Alice, a 27-year-old engineer, told me: 'When I'm leading a team, if there's a guy in exactly the same position, I get questioned a lot more. I have to fight more for the same sense of authority. My experience counts, but not as much as a man.'

Spotting our own biases is a start, but it's not the end. We need to address the problem at a structural level too. As long as we see many more men than women in positions of authority, we will tend to associate men with authority and women with subordinate status. As long as we allow boys to grow up believing that they are superior to girls, we are instilling habits of mind that will be very hard to change in later life. As long as we keep women in the workplace down by punishing them for being as assertive or self-promoting as men, they will never advance in the same numbers. And as long as we make work patterns unfriendly for parents of both genders, we are going to prevent women from reaching the positions of authority they need to for society to rebalance its stereotypes.

In some countries, the bias is not remotely unconscious but runs visibly through the veins of the whole society. The documentary-maker Sharmeen Obaid-Chinoy has won two Oscars for her films highlighting the lowly status of women in Pakistan. 'Pakistan is a highly misogynistic country. It's deeply patriarchal,' she told me. 'Whether it's in small towns, villages or big cities, by and large men make decisions for women. Very few women are allowed to make decisions about their own lives. You will find the most educated women, the most empowered women on paper, unable to make simple decisions about their lives.' And if they rebel against this? 'The women who have the courage to articulate, to ask, to push, to demand more rights are maligned, are labelled and are often killed. So very few women now dare to publicly do that.'[11]

In the developed world, thankfully, women are usually allowed to make decisions about their own lives. They can and do speak out for their rights without having to fear for their lives. But that doesn't mean that the problem has been solved, for covert sexism is very hard to fight. It is far easier for perpetrators to deny or dismiss. Women who complain about instances of it can be caricatured as chippy, over-sensitive or humourless, or told they are being hyster-ical and making it up. In the cases of women such as Mary McAleese, Frances Morris and Elaine Chao, I somehow doubt it. Their stories give credence to those of the rest of us.

What do we mean by 'authority'? I am using two definitions. The first is the influence that people have as a result of their knowledge and expertise – in other words, being considered authoritative on a subject.

The second is the exertion of power and leadership – in other words, having authority as a result of being in charge. This could as easily refer to authority within a family as to authority in the public sphere, but in this book I am writing about authority outside the home. Women throughout history have too often been palmed off with being given charge of the household. That's fine for some, but for many of us it's not enough. We deserve to be treated equally in all walks of life.

So how can an authority gap persist in the twenty-first century, when we've had two female prime ministers, and women as President of the Supreme Court, First Ministers of Scotland and Northern Ireland, and Commissioner of the Metropolitan Police? If authority is wrapped up with power and leadership, shouldn't we be over this by now? And if not, isn't it just a phenomenon of the older generation who were brought up with values formed in the 1960s and 1970s, when women still gushed in TV ads about Persil washing whiter and men in suits bossed their secretaries around?

Surprisingly, perhaps, we're still at it, even the young. I've talked to female university students who are clearly as clever as their male counterparts. Yet even they find themselves patronized and underestimated. Flora, 20, says, 'I have to work harder than the males in the group do to put my point across. Especially in a group that is male-dominated, people tend to listen first to the men. If a man tries to take charge in a group scenario, they are often taken more seriously. When female friends make suggestions, they are often shut down by the men making jokey comments which encourages other members of the group to chime in.'[12]

Ellie, 20, adds: 'In my subject group the men prefer to work together and exclude the women from interacting with them academically. There is an assumption that we are not their intellectual equals.'[13]

The trouble is, changes in the outside world take a long time to percolate into our subconscious, to change the stereotypes that lurk there. Until then, we have to actively notice when our unconscious bias plays tricks on our thoughts about other people – and that isn't easy. As Julia Gillard, former Prime Minister of Australia, who has co-written a book about women and leadership, told me: 'We've never lived in an environment free of any stereotyping. We've never lived with true gender equality. So even young people who are very highly sensitized, and don't want to be discriminating on the basis of gender or indeed on the basis of anything else, can't shed all of that social conditioning just through an act of will. You can't say, "I'm a feminist," and somehow all your social conditioning is gone. We need to be second-guessing what's happening in our brains, so we don't just give in to it.'[14]

But why does this happen? Well, millennia of men being in charge, of the patriarchy, have made their mark on our minds. They have made it quite normal for men to condescend to women and to treat them as subordinate. As recently as the early twentieth century, the novelist Arnold Bennett wrote a book called *Our Women* which included a chapter entitled 'Are Men Superior to Women?'[15] It's not clear why he bothered to use a question mark. For in it he wrote: 'The truth is that intellectually and creatively man is the superior of women, and that in the region of creative intellect there are things that men almost habitually do but which women have not done and give practically no sign of ever being able to do.' No wonder the book enraged Virginia Woolf, who wrote two letters to the *New Statesman* taking issue with it.[16]

Aiming her barbs at the *Statesman*'s literary editor, Desmond MacCarthy, who had reviewed Bennett's book under the *nom de plume* of Affable Hawk, Woolf wrote:

The fact that women are inferior to men in intellectual power, he says, 'stares him in the face.' He goes on to agree with Mr Bennett's conclusion that 'no amount of education and liberty of action will sensibly alter it.' How, then, does Affable Hawk account for the fact which stares me, and I should have thought any other impartial observer, in the face, that the seventeenth century produced more remarkable women than the sixteenth, the eighteenth than the seventeenth, and the nineteenth than all three put together? . . . In short, though the pessimism about the other sex is always delightful and invigorating, it seems a little sanguine of Mr Bennett and Affable Hawk to indulge in it with such certainty on the evidence before them. Thus, though women have every reason to hope the intellect of the male sex is steadily diminishing, it would be unwise, until they have more evidence than the great war and the great peace supply, to announce it as a fact.

Even among those with average intellect, Bennett claimed that men were still superior: 'Every man knows in his heart, and every woman knows in her heart, that the average man has more intellectual power than the average woman . . . It is a fact as notorious as the fact that a man has more physical strength than a woman.' This book was published in the 1920s, the same decade as my mother was born: a woman who won places at both Oxford and Cambridge at the age of sixteen, despite having missed out on several years of schooling thanks to the Second World War.

It would be a foolish man who dared to write such nonsense today, in the face of scientific evidence showing women to be absolutely the intellectual equal of men. But we still had V. S. Naipaul, the novelist, claiming in 2011 that no female writer had ever written as well as he.[17] And the American writer Norman Mailer writing in 1959 that 'a good novelist can do without anything but . . . his balls'.[18]

And the problem is that the world around us is still designed and led mainly by men. Most of us have grown up with our fathers working more and earning more than our mothers. We see, in all walks of public life, men making it more often to the top, and men being cited as authorities much more often than women. We have watched too many films in which men are the protagonists and women the helpmeets or sex objects, in which men have twice as many speaking parts, almost all of which are directed by men.[19] We still live in a world in which men have the upper hand and help each other up the ladder, so no wonder we have internalized the notion that women must somehow be inferior and worthy of less respect.

Mahzarin Banaji is Professor and Chair of the Department of Psychology at Harvard and an expert on unconscious or 'implicit' bias. This, as she told me, is how it comes about: 'Implicit bias comes from our social world, from our culture, because the content of what the brain knows is what it sees in the world. So I see that men do certain kinds of work and women do other kinds of work. If I had seen in my world that women were largely construction workers and engineers, that's what my brain would have learnt; and if I had seen in my world that men largely took care of children at home and cooked and cleaned for them, then that's what my brain would have learnt.'[20]

We absorb the notion of male superiority from such an early age. British parents, when asked to estimate their children's IQ, will put their son, on average, at 115 (which in itself is hilarious, as the average ought to be 100) and their daughter at 107, a huge statistical difference.[21] Why they do this is a mystery, as young girls develop faster than boys, have a bigger vocabulary, and do better at school. But the result is that boys, on average, grow up thinking that they are cleverer than girls, and vice versa. As early as five, studies show that children believe girls aren't as good at maths as boys (even though they are).[22] And when asked to choose team-mates for a game for 'really, really smart' children, young children of both genders are more likely to choose boys than girls.[23] Yet, at that age, girls are ahead of boys academically, and the children in the study knew it.

American parents, meanwhile, are two and a half times more likely to Google 'Is my son gifted?' than 'Is my daughter gifted?' even though girls make up 11 per cent more of the gifted and talented programmes in US schools.[24]

No surprise, then, that adult men will, on average, put their own IQ at 110, while women estimate themselves to be just 105.[25] Yet we know that, except for at the far extremes of the IQ curve, women's and men's IQs are distributed identically. Girls, on average, get higher grades at school and are more likely than boys to win places at university, masters courses and beyond. The main reason for boys and men to think that they are cleverer than girls and women must be because their parents, teachers and society have – incorrectly – imbued them with the belief that they are.

Some readers will claim that this is all old hat. Aren't women now being favoured? In fact, aren't they now getting all the top jobs? Isn't it hardest of all these days to be a middle-aged white man: 'pale, male and stale'? Well, it *is* harder than it was, when all those characteristics conferred a massive advantage in every aspect of life. And it's true that some of the top jobs that have been held by men throughout their existence are at last having women appointed to them. Structurally, things are starting to change for the better. But, as the experience of the highly successful women I've quoted shows, even getting a top job doesn't entirely insulate you from having your authority challenged.

Nor is the change happening so fast that men are unfairly suffering, though it may feel that way when male privilege starts to be withdrawn. And by 'privilege' I don't mean wealth or social status: merely the fact of being a man rather than a woman. Boris Johnson, for instance, was praised for promoting women in his first Cabinet, but they still occupied only eight out of thirty-three positions: in other words, there were three men for every woman around that table, in a supposedly representative democracy. And, at the time of writing, there are still only six female CEOs of Britain's hundred biggest listed companies. Women still have a very long way to go – and men still have a very long way to fall – before we get anywhere near equality.

Yet, just as men tend to feel that women are dominating a discussion when they talk for only 30 per cent of the time, they often feel that women are getting an unfair crack when they are merely, at last, being treated a little more equally.[26] A former editor of mine told me that my book was out of date because the only people being appointed to boards these days were women; men no longer had a chance. The next day, I sent him the figures for the previous month: there had been twenty male board appointments and nineteen female ones. Nothing like enough to even out the existing 2:1 ratio of men to women on boards, but better than it used to be. And although board appointments are now, at last, more equal thanks to government insistence, the same isn't true for executive jobs. For him, though, it felt like male annihilation.

As the philosopher Kate Manne puts it in *Down Girl*: 'These bastions [of privilege] are often well-defended and difficult to challenge. For people are often, unsurprisingly, deeply invested in their continuation. To make matters worse, these structures are often quite invisible to the people whose privileged social positions they serve to uphold and buttress. So dismantling them may feel not only like a *comedown*, but also an *injustice*, to the privileged. They will tend to feel *flattened*, rather than merely *levelled*, in the process.'[27]

She is right about the invisible nature of privilege, which is the flip side of bias. Most men simply don't notice it. And why would they? I struggle to notice my white privilege, the fact that people aren't biased against me because of my skin colour. Yet, in everyday life, it's as if men are swimming with the current in a river and women are swimming against it. The men see the banks racing past them and congratulate themselves for swimming so powerfully. They look at the women struggling to make headway against the current and think, 'Why can't they swim as fast as me? They're obviously not as good.' Unless they make a conscious and sustained effort to do so, men can't feel the current, and they put their success – and women's relative lack of it – down to pure merit. It's human nature not to want to believe that privilege, and the bias it engenders, has helped them along. Or that women are being held back, despite their merit.

This means there is a deep asymmetry. When I talk to men about this subject, some of them – like my former editor – express scepticism. They tell me that the problem has been solved, that my thesis is out of date, because lots of women are being appointed to top jobs. If anything, women are being favoured and men are suffering. This is because they can't see the continuing bias and we can. They don't experience the myriad little insults to their self-esteem and competence that women have to put up with daily or weekly. Not only is their reaction as wrong-headed as a white person telling a person of colour that racism doesn't exist; it also proves the exact point of this book, that women's authority is questioned and challenged even when they know more than the person they are talking to. For, as you will see, there is a huge amount of evidence for the continued existence of the authority gap. It's resistant to being mansplained away.

Although there has been some recent progress at the very top, with women being appointed to jobs that have always been held by men, the underlying assumptions about women's expertise don't seem to have changed much. When I interviewed her, Liz Truss was Chief Secretary to the Treasury, but after Boris Johnson took over as Prime Minister she had made it no secret that she wanted to be Chancellor of the Exchequer (a job no woman has held). A lot of people questioned whether she would be up to it, even though her background was in economics and finance. She is gloriously stroppy about the way she is still treated. 'As a woman in politics you have to prove your expertise more,' she told me.[28] 'So, I'm a qualified accountant and an economist and I'm Chief Secretary to the Treasury [the Chancellor's deputy], but there are still question marks. I got asked how good my maths was the other day, for God's sake. I've got a double A level in maths!' I don't recall people presuming that her former colleague, George Osborne, wasn't up to the job of Chancellor, even though he read history at university and had never done a job involving finance or economics.

The same still happens in the private sector too. Anne Mulcahy – tall, commanding and super-articulate – is an extraordinarily successful American businesswoman. She took over as CEO of Xerox when it was a giant but struggling photocopier company being superseded by digital technology. It looked destined to go the way of Kodak. The day she was appointed, the shares fell 15 per cent, which was, as she wryly said, 'a real confidence-builder'. The company was on the verge of bankruptcy, had been making losses for six years and had over $17 billion in debt. After Mulcahy brought it back to profitability against all the odds, *Money* magazine described it as 'the great turnaround story of the post-crash era'.[29] And, in 2005, *Fortune* magazine named her the second most powerful woman in business, after Meg Whitman, CEO of eBay.

Given this, you might think she would be seen as an unquestioned business superstar. She certainly would if she were a man with the same record. Yet, as she told me, 'When I join a new board, when I take a new role, there's a "wait and see". You don't come in with that kind of assumption that you've earned where you are and therefore you don't have to prove yourself again.'[30]

These accounts may be anecdotal. But the experience of having our authority and expertise underestimated or challenged is something that pretty much all women share. Research shows that women are twice as likely as men to say that they have to provide more evidence of their competence. They are much more likely than men to say that their expertise is challenged, that they are interrupted or talked over and that other people take credit for their ideas.[31] When I was writing this book and told people what it was about, some men seemed perplexed, but every woman was delighted. All smiled in recognition, and many said, 'Yesss!' and that they couldn't wait till the book came out.

But we don't just have to ask ourselves and other women about it. For this book, I have scoured the academic and professional worlds for concrete evidence and studies on women's authority, influence, competence and power. I am going to bring you research from a wide variety of disciplines – psychology, sociology, linguistics, politics, management and business – and I have also commissioned new research. I have done my best to suggest solutions: ways for us to counter our bias and to bring up a new generation to think and behave differently. And I want to show what we can all gain – men as well as women – from taking women more seriously.

I

You don't have to read this chapter

(unless you're a sceptic about the authority gap)

'We did not like to declare ourselves women, because . . .
we had a vague impression that authoresses are liable to
be looked on with prejudice.'

– Charlotte Brontë

M OST WOMEN WILL instantly recognize the phenomenon of the authority gap. It's as obvious – and as annoying – as the gender pay gap. More enlightened men appreciate it too. If you're one of the above, feel free to skip to the next chapter. But if you're sceptical that the gap exists, or if you secretly believe that it may be justified, please read on for the irrefutable evidence that supports my case.

When the Boston writer Catherine Nichols finished the final draft of her first novel, she asked several accomplished writer friends to have a look at it. They reassured her that it was ready to go, and that she should start approaching agents. So she sent the same covering letter and first few pages of the novel to fifty agents and sat, excited, to see what the response would be.

She waited . . . and waited . . . and waited. Eventually the replies started trickling in: all rejections. Out of the fifty agents, only two asked to see the full manuscript, but with no guarantee that they would be interested in representing her. 'My writer friends still promised it was a good book, that I should have faith in my work, that good news would be around the corner. It wasn't.'[1]

So Nichols conceived what she called her 'nutty plan'.[2] She created an email account in a man's name, similar to her own, and decided to send exactly the same letter and sample chapter to some more agents. She sent off the first submission, and before she had even drafted the second she received a reply saying, 'Delighted. Excited. Please send the manuscript.'[3]

To the six queries she sent on the first day, she received five instant replies. Three asked for the manuscript and two were warm rejections, praising his 'exciting' project. 'It was shocking how fast it became obvious there was a big difference,' she told me.[4] So she decided to approach fifty agents under the male name, just to see exactly how big that difference was. She got seventeen positive replies compared with two, which meant, she joked, 'I was an eight and a half times better writer under a male name.' One agent, who had rejected her as Catherine, not only asked to read George's book, but wanted to send it to a more senior colleague.

Even the rejections were more complimentary and helpful. As a woman, she had won praise for her 'beautiful writing', but that was it. As a man, she was told her work was 'clever', 'well constructed' and 'exciting' and was given useful advice as to how to improve the novel. 'That is where I went from feeling flattered to feeling angry,' she told me. 'As a man, I felt the critical responses were getting into the structure of the book, or the thought processes of the characters, or the mechanics of the plot. That level of attention was truly helpful when it came to redoing parts of the book.' As a man, she was being coached, but not as a woman. And many more of these agents were women than men.

'So what must it feel like to be a man?' I asked her. She burst out laughing. 'It must feel *amazing* to be a man!'

Nichols had changed only the crucial variable: gender. Her experience may be anecdotal, but social scientists have replicated it in rather more rigorous experiments. Corinne Moss-Racusin, John Dovidio and other colleagues, in a randomized double-blind study, sent out applications for a lab manager position to male and female science professors at top universities.[5] The application materials were identical, but the applicants were randomly assigned male or female names.

The professors rated the 'male' applicant as significantly more competent and hireable than the (identical) 'female' one. They offered him a higher starting salary and more career mentoring. The stereotypes in these professors' brains were playing havoc with their rational, scientific judgement. As a result, the professors conferred more authority and expertise on the 'male' applicant than the 'female' one, even though the two were identically qualified. Their unconscious bias created an authority gap where none should have existed.

What is more, the female professors were just as biased as the male ones. We'll see in Chapter 9 why that is. Dovidio, a Professor of Psychology at Yale, told me, 'That tends to perpetuate it even more, because if you have people making a joint decision and a member of the minority group goes along with members of the majority group who have these subtle biases, it legitimizes the response, where you're going to continue to hire the man rather than the woman. When women exhibit this subtle bias, it frees up men to exhibit this subtle bias more.'[6]

Another study asked students taking an online course to rate their instructor.[7] If they thought they were being taught by a man, they gave higher ratings than if they thought they were being taught by a woman. The instructor, of course, was the same person. According to the authors, 'The results were astonishing. Students gave professors they thought were male much higher evaluations across the board than they did professors they thought were female, regardless of what gender the professors actually were.' So the 'men' were being accorded greater authority than the 'women'.

A similar study with psychology professors – both male and female – found that they rated identical CVs more highly when they had a man's name at the top, and were more likely to offer him a job.[8] They were also more likely to say that he had good research and teaching experience, even though the applicants' experience was exactly the same. Interestingly, though, when the professors were sent a more senior scientist's CV and were asked whether 'he' or 'she' should be awarded tenure, there seemed to be no bias against the female applicant. However, the authors acknowledge that they probably sent too good a CV for this part of the experiment, and it would have been hard to resist recommending tenure. They conclude: 'A superb record may indeed function as a buffer for gender bias when making promotional decisions.' That explains why outstanding women sometimes do manage to make it to the top in real life, albeit in smaller proportions than men do.

But even though the exceptional 'female' applicant was recommended for tenure as often as the 'male', the professors were four times more likely to write cautionary comments in the margins of their responses to her, such as, 'It is impossible to make such a judgement without teaching evaluations'; 'We would have to see her job talk'; and 'I would need to see evidence that she had gotten these grants and publications on her own.' It's extraordinary, and depressing, that – even in the twenty-first century – people still find it hard to believe that women can be exceptional. On the 'male' application, by contrast, there were few such cautionary comments.

A 'superb' record of achievement can definitely act as a buffer against bias. Many of the extremely successful women I've talked to during the research for this book had this record. And they knew that being really, really good at their job was their only hope of advancement. As the Chinese-American businesswoman Wan Ling Martello put it to me: 'More than knowing what you're doing, you have to be super competent at what you're doing. That is to me first and foremost. That just gives you licence to play with the boys. Without that, you can forget it.'[9] For women in professional jobs know that they can't afford to fail; unlike men, they are unlikely to be given a second chance.

Once my interviewees got to the very top, some of them said they felt much more protected from overt instances of the authority gap. It was harder for people to disrespect them or discount their opinions once they were in charge.

I went to visit Helle Thorning-Schmidt in the north London home that she shares with her British MP husband, Stephen Kinnock. She's fizzing with energy and talks faster in English than most of us native speakers. She told me: 'There's a huge difference between getting to the top and being at the top in terms of how you get treated as a woman, and I really felt that once I was Prime Minister that I didn't get treated very differently from men. It was because, being Prime Minister, I had the authority and I had a lot of power.'[10] She was running a country in Scandinavia, though, which was already well in advance of most other nations in its attitudes to gender equality. As we'll see in Chapter 3, the female leaders of Chile and Australia didn't find themselves totally immune from the authority gap even when they reached the pinnacle of power.

The world of business may be better for female leaders once they finally reach the C-suite. Anne Mulcahy suffered some horrible sexism on her way up the corporate pyramid to run Xerox. But 'when you become CEO, it's almost like you're an honorary man. There's power that comes with the position, so it's harder for people to confront or not listen or not be respectful. That's fabulous, right?'[11]

It sure is. But without the insulation afforded by being the ultimate boss (and sometimes even with it), women are still thought to be less authoritative than men. We seem to be programmed to assume that men are more expert and more hireable than women.

Could there perhaps be a genuine justification for this bias? Could it be that, on average, men are indeed cleverer, better qualified and better informed than women, and that it therefore makes sense to rate them more highly and respect them more? In other words, is the authority gap justified? Let's look at the evidence. From a very early age, girls outperform boys. They develop faster, talking earlier, learning self-discipline at an earlier age, and using a bigger vocabulary.[12] They get better grades at school, particularly in the humanities, but also in maths and science in some countries, and outnumber boys at university.[13] [14] In the US, they win 57 per cent of masters degrees and 53 per cent of doctorates.[15] So there is no question that they are better educated and qualified than men, at least in the younger generations, who weren't held back from going to university.

On average, girls and women are exactly as intelligent as boys and men.[16] There *is* evidence suggesting that boys and men have more variable IQs: that there are a lot more males at the very bottom of the IQ distribution and a few more at the very top.[17] The disproportion at the bottom is far higher than that at the top: those with the very lowest IQs tend to be male, as boys are more prone to developmental disorders. It's not clear whether the smaller differences at the very top are due to biological or social differences. As we saw earlier, parents tend to believe their sons are cleverer and may encourage them more. And as early as the age of six, children of both genders believe that boys are more likely to be 'really, really smart' than girls, which probably gives girls less intellectual confidence and boys more.[18]

Is there any evidence that can help us decide whether this outperformance at the very top end of the intelligence scale is a biological difference between the genders or something that reflects the way we are brought up? It is interesting that, back in the 1970s and 1980s in the US, boys who were *exceptionally* good at maths outnumbered girls by thirteen to one. Now the ratio is as low as two to one.[19] In some countries and among some ethnic groups, the gap doesn't exist.[20] This strongly suggests that the difference in mathematical ability is unlikely to be biological or innate, or it wouldn't vary across time, place and ethnicity. Indeed, in countries with greater gender equality, there are more gifted and profoundly gifted female mathematicians.[21] So it seems likely to be more a matter of expectations – from parents, teachers and other children – and social conditioning than anything inherently different in our brains.

What is certainly true is that at the very top levels of academic achievement at school, girls are easily outperforming boys. In 2019, the UK brought in a new system of grading GCSE exams, which went from 9 at the top to 1 at the bottom. A 9 is an even higher grade than the previous A*, and very few students were expected to win a clean sweep of 9s. In the end, 837 pupils in the whole country did so, representing the top 0.1 per cent of exam-takers. Of these, girls outnumbered boys by two to one.[22] So there is absolutely no evidence on which to base the lazy assumption that boys or men are cleverer than girls or women: quite the opposite.

People still worry that not enough girls study STEM subjects, and many secretly suspect that girls' brains just aren't cut out for maths and science. But there is surprisingly little angst about the consistent underperformance of boys in humanities. In literacy, fifteen-year-old girls are significantly better than boys in every single one of the seventy-two countries surveyed by the OECD's Programme for International Student Assessment.[23] In these countries, girls, on average, match boys at maths and are only marginally behind in science, and they are well ahead at reading. So when it comes to choosing what to study at A level or for university, and teachers say to students, 'Do what you're best at,' it's quite logical that boys have been more likely to choose STEM subjects and girls humanities. It's a question of comparative advantage, not of girls' brains being unsuited to science. According to one study, it explains up to 80 per cent of the gender gap in choosing maths-related studies and careers.[24] But even this is beginning to change. In 2019, for the first time, girls marginally outnumbered boys in the UK in taking science A levels, perhaps because schools are at last encouraging them to do so.[25]

The one science in which girls still lag well behind is physics. Boys outnumber girls taking physics A level by more than three to one. Interestingly, girls at single-sex independent schools are four times more likely to do A-level physics than girls at co-ed state schools, which again suggests that this is a social rather than an intellectual problem.[26] Girls may feel less comfortable studying what's seen as a rather masculine, nerdy subject alongside boys in the classroom.

The same is true for young women at university. In one study, female students were asked why they weren't majoring in subjects such as engineering and computer science.[27] It wasn't the maths or science that was putting them off or anxiety that the subjects might be too difficult. Overwhelmingly, it was that they thought there was gender discrimination on these courses.

And they'd probably be right. A study of female graduate students in physics and astronomy found that 76 per cent said they had suffered sexism, and this included real evidence of the authority gap.[28] For instance, Janet told the researchers, 'I feel that I am not listened to within my group. This is mostly by my peers (my adviser tends to do a better job). A lot of my suggestions are brushed off. If, later, they turned out to be correct, people forget that I ever made them. In another situation, I told the grad student under me that he should consider a certain factor in trying to make sense of his data. He said no and ignored me. When the other [male] grad student/post doc suggested it, he was open to it right away.' She's describing the authority gap in action.

If women's abilities are every bit as good as men's, maybe the problem is that they don't have the personality to earn authority? Psychologists Paul Costa, Antonio Terracciano and Robert McCrae compared the scores of 1,000 American men and women on a personality test designed to assess the 'big five' personality traits: extraversion, agreeableness, openness, conscientiousness and neuroticism.[29] The gender differences were not the same across all cultures, suggesting that these characteristics are not evolutionally determined, as some psychologists would have us think. And the two traits that are best correlated with effective leadership – extraversion and openness – exist at roughly the same level in women and men.

Women score more highly than men on warmth, positive emotions, gregariousness and activity. Men score more highly than women on assertiveness and excitement-seeking. But the differences aren't great and neither gender has an overall advantage. Men might make better leaders if they were warmer and more positive; women if they were more assertive.

Women's assertiveness has increased over time, suggesting again that it's not a question of hormones, evolution or neurological difference.[30] Compared with how women judged themselves on the same qualities decades ago, today's women say that they are more ambitious, more self-reliant and more assertive but they have not lost traditionally feminine traits, such as affectionateness or understanding. Men, meanwhile, show little change over time. In many studies now, the sex differences in assertiveness are quite small.[31]

Men seem to be more likely to take risks than women, though the difference is also not great and has been getting smaller over time.[32] Besides, risk-taking is hardly an unqualified asset in a leader – just look at the years leading up to the financial crisis or the botched management of the coronavirus pandemic by blustering leaders such as Boris Johnson, Donald Trump or Jair Bolsonaro. Female leaders performed much better.

Empathy and emotional intelligence, which are important for good management, seem to be stronger in women than men.[33] Women are also less tolerant of dishonest or illegal behaviour, such as taking bribes.[34] But what about competitiveness, which evolutionary psychologists claim is embedded more in men than in women because men had to compete for mates by gaining more strength, status and power?

If competitiveness were an innate, biological difference between women and men, then it would remain the same across cultures and across time. But researchers who studied the matrilineal Khasi society in India and the patriarchal Maasai society in Tanzania made some very interesting discoveries.[35] The participants were asked to throw a tennis ball into a bucket but could choose whether to be rewarded with a small amount for each successful shot, or three times as much if they scored better than an opponent. Maasai men proved to be twice as competitive as Maasai women, but Khasi women were twice as competitive as Khasi men. And Khasi women were *more* competitive than Maasai men.

Context is all. Most modern societies are patriarchal, so men tend to be more overtly competitive than women (though we are only comparing averages here, not individual men and women, who may differ greatly from the average). But even here, the average differences are very small, and show only slightly greater male competitiveness and slightly greater female cooperativeness. Many women – like me, I have to admit – are pretty competitive. (Did my admission make you feel momentarily uncomfortable? Would it have done so if I'd been a man?)

Perhaps, then, women are less ambitious than men? It's certainly true that some women put family before career, and more do so than men. However, not all of this behaviour is entirely voluntary. Because the gender pay gap starts before couples have children, when a woman becomes pregnant it often makes more sense for her to dial back her work, or give up altogether, if her partner is the one who is more highly paid. And because women still do so much more of the unpaid work at home – 60 per cent more than men in the UK – holding down a demanding job while bringing up children can be a daunting and exhausting task.

Women and men who have already reached senior levels of management say they have similar ambitions to get to the top.[36] Yet it's almost always the men who get there. This can't be entirely on merit, since women's performance evaluations, on average, are every bit as good as men's. So if it's not due to less ambition and it's not due to merit, what is left? Only unconscious or conscious bias.

Even as university students, young women are as ambitious as young men, although women are more pessimistic about their chances of achieving their ambition.[37] Their pessimism may be justified. Catalyst surveyed nearly 10,000 MBA graduates from leading business schools in the US, Canada, Europe and Asia.[38] It found that, even among those graduates who aimed to become CEOs or senior executives, the men were offered better jobs with higher salaries straight after graduation. The researchers controlled for experience, industry and global region and found the same results. This was nothing to do with parenthood, as the gap was the same for childless men and women.

Maybe employers simply don't want to take the risk that a woman in her twenties or thirties will leave her job to start a family? I'm sure this is part of the problem for women, but it's based on a false premise. In reality, there are many reasons to leave a job, such as going for a better one with a different employer, and it turns out that men and women leave their jobs at the same rate,[39] and in fact male managers are *more* likely to leave than female managers.[40]

Yet across the board – not just for MBA graduates – studies show that women are promoted more slowly than men with the same level of education and experience. After being hired, they have to wait longer than men to be made a supervisor or a manager, and then they have to wait longer between promotions.[41] So they are not being awarded the authority they deserve.

A huge survey, *Women in the Workplace 2020*, which looks at nearly 600 companies and questions a quarter of a million employees, finds that for every 100 men promoted to manager, only 85 women are promoted – and this gap is even larger for women of colour: only 58 black women and 71 Latinas are promoted for every 100 men.[42] As a result, in 2020, women held just 38 per cent of manager positions, while men held 62 per cent. And this is true even in female-dominated fields. For instance, in the UK, women make up 64 per cent of secondary-school teachers but only 39 per cent of headteachers.[43] What's more, female headteachers earn, on average, 13 per cent less than male ones.[44] So we can't just fall back on the excuse that women are paid and promoted less in traditionally male-dominated fields, such as STEM, because they are going against stereotype. When men go against stereotype, for instance by going into teaching, they scythe their way through the ranks. There must be sexism at play.

If women do make it to the top, despite the obstacles, do they make as good leaders as men? The evidence shows that they are even better. A meta-analysis of 99 studies found that women were rated by other people as significantly more effective leaders than men, although male leaders rated *themselves* more highly than women did.[45]

Women are particularly good at people management. They are better than men at what is known as 'transformational' leadership, in which they mentor and empower employees, encourage them to develop their full potential, engage their trust, and allow them to contribute their views – in other words, being democratic rather than autocratic leaders.[46] They are also more likely to use rewards for good performance. By contrast, men are more likely to be *laissez-faire* managers, to wait for problems to become serious before intervening, and to focus on employees' mistakes. These styles don't promote effective leadership, while the transformational and rewards styles do.

We all saw, during the Covid-19 pandemic, how much more successful, on average, female leaders were than male ones. Globally, women held only 7 per cent of government leadership positions, yet four of the top ten countries which dealt best with Covid-19 were led by a woman.[47] Nicholas Kristof of the *New York Times* looked at death rates in thirteen countries led by men and eight led by women after the first few months of the pandemic.[48] The male-led ones had an average of 214 coronavirus deaths per million; the female-led ones had only 36 per million.

As he writes: 'It's not that the leaders who best managed the virus were all women. But those who bungled the response were *all* men, and mostly a particular type: authoritarian, vainglorious and blustering. Think of Boris Johnson in Britain, Jair Bolsonaro in Brazil, Ayatollah Ali Khamenei in Iran and Donald Trump in the United States. Virtually every country that has experienced coronavirus mortality at a rate of more than 150 per million inhabitants is male-led.'

He believed that it was a question of ego. Female leaders were happy to put public health first, to defer to scientific experts and to act quickly. Most of all, they were humble. They didn't assume that their country would be fine because either they or the nation were somehow exceptional. There was no bluster or strutting; they just got down to the job.

A more rigorous study of this phenomenon was conducted by two economists, Supriya Garikipati and Uma Kambhampati, for the World Economic Forum.[49] They compared countries led by women with other nations that had similar demographics and economies. So New Zealand was matched with Ireland; Bangladesh with Pakistan; Germany with the UK. Still they found that the female-led countries locked down earlier and had significantly fewer cases and deaths than the male-led ones. The results were 'especially highly significant' on deaths.

Could this be because the female leaders were more risk-averse? Yes and no. They clearly wanted to take fewer risks with their citizens' lives, but they were prepared to take more risks with their economies. The authors conclude that it was a combination of good communication skills and transformational leadership that led to these hugely better results.

Almost all the senior women I interviewed for this book use transformational leadership, which – incidentally – often helps to defuse hostility to a female boss. It can melt the resistance to female authority by tempering the necessary decisiveness of a leader with the warmth that people expect – and demand – from women.

Muriel Bowser, the Democrat African-American Mayor of Washington DC, is the sort of woman you would happily allow to lead you out of a burning building. She exudes calm and competence. When I visited her in her office, we talked about her leadership style. 'There's the formal power of the office and there is the power that you get from building relationships, having been in the trenches before and working together. There are so many more things that I've been able to control through the informal leadership techniques that I've developed over the years. Direct communications, phone calls, text messages, having lunches with people, thanking them when something good happens, supporting them when something not so good happens.'[50]

And here's Elaine Chao, Donald Trump's Transportation Secretary: 'I think I'm very inclusive, and very participatory. I constantly ask for feedback. And I'm not afraid of people who are smarter than I am. I want people who are smarter than I am. The teamwork is very important.'[51]

Baroness (Eliza) Manningham-Buller was only the second woman to run the British security intelligence agency, MI5. She is a woman for whom the adjective 'formidable' could have been invented. But she recognizes the importance of humanity in leadership and thinks the conventional wisdom is often mistaken. 'Books on leadership often pick on the wrong things. They talk about your ambition, your targets, your mission, your objectives, your project management, all that stuff. And that's all important. But how you behave is more important, because you must have a degree of humility, you must treat people decently and fairly, you should expect high standards of them and demand that they fulfil them, but in doing the job you should have fun, humour, compassion, the softer things, because that's why people stay in an organization. They feel comfortable and they feel their voice is heard and they are not just a cog.'[52] Female leaders are often more comfortable expressing humility, which in a man might be thought to be weak. And their subordinates generally appreciate it.

Janet Yellen, US Treasury Secretary, has a similar philosophy: 'I have worked in organizations where the pay is not why people are there. They want to feel they're making a contribution, really doing good work, and they're pouring their heart and soul into what they have done. And I'm very aware that saying, "Thank you, that was really great what you did, I really appreciate it, you did a fine job," that means a ton to people.'[53] Employees often say that feedback is one of the most important things they want from a boss. As we'll learn later, this is one of the reasons why both sexes are more engaged when working for a woman.

So what does the evidence tell us? Women are just as intelligent and, if anything, they are better educated than men. They are just as competent in their jobs. Those who stay on the career ladder are just as ambitious as men. And if they get to the top, they are – on average – better leaders than men. Yet they don't win the rewards they deserve for their intelligence, education and competence because we still, instinctively, underestimate women's worth, challenge them more and promote them less. We still, deep down, act as if men are cleverer than women, more deserving of good jobs, and more worthy of respect. And that's what creates an authority gap.

2

The view from the other side

What we can learn about men and women from people who've lived as both

'Men treated me more and more as a junior . . . and so, addressed every day of my life as an inferior, involuntarily, I accepted the condition. I discovered that even now, men prefer women to be less informed, less able, less talkative and certainly less self-centred than they are themselves: so I generally obliged them.'

– Jan Morris, author and trans woman

I F A WOMAN thinks that she's being taken less seriously than a male rival, it's hard for her to prove that she's the victim of discrimination. Women are often accused of citing sexism to disguise their own inadequacy. But there's one very persuasive way of testing the hypothesis that women are, unjustifiably, respected less and thought to be less competent and expert than men: talk to people who've lived as both. Because they are exactly the same person – with the same ability, experience and personality – and the only thing that has changed is their gender, they are uniquely able to identify the effect of gender in their lives. It's a way of correcting for all the other variables and isolating the one that matters.

When Ben Barres was a student at the Massachusetts Institute of Technology, then living as a woman, his maths professor set the class a demanding test with five questions. The last one was particularly hard, but Barres still managed to solve it. The following day, the professor handed back the papers and announced that no one had succeeded in answering the fifth question.

Barres picks up the story: 'I went to the professor and I said, "I solved it." He looked at me and he had a look of disdain in his eyes, and he said, "You must have had your boyfriend solve it." I didn't know what to say. He was in essence accusing me of cheating. I was incensed by that.'[1]

For the next few decades of his career as an academic research scientist, before he transitioned, Barres found himself at a disadvantage. Much later, he wrote: 'I am still disappointed about the prestigious fellowship competition I later lost to a male contemporary when I was a PhD student, even though the Harvard dean who had read both applications assured me that my application was much stronger (I had published six high-impact papers whereas my male competitor had published only one).'[2]

Then, as a middle-aged Stanford professor of neuroscience, he transitioned and became known as Ben Barres. He was astonished by the difference it made to his life. 'I've had the thought a million times,' he wrote, 'I am *taken more seriously*.'[3] At one seminar, a faculty member who didn't know his history was heard to say, 'Ben Barres gave a great seminar today, but then his work is much better than his sister's.'[4] Barres concluded: 'By far the main difference that I have noticed is that people who don't know I am transgendered treat me with much more respect: I can even complete a whole sentence without being interrupted by a man.'[5]

Because Barres, who developed baldness and grew a beard, began being treated as one of the guys, he started to hear male colleagues say what they really thought about women – things they would never have said to him before he transitioned because they revealed the men's sexist attitudes. 'A neurosurgeon at Stanford told me he'd never met a female neurosurgeon who was remotely as good as a man. Another told me he thinks women are like small children.'

By coincidence, another middle-aged science professor at the same university was transitioning at the same time from male to female, to be known as Joan Roughgarden. 'It was clear when I got the job at Stanford that it was like being on a conveyor belt,' Roughgarden said of her time living as a man.[6] 'The career track is set up for young men. You are assumed to be competent unless revealed otherwise. You can speak, and people will pause and people will listen. You can enunciate in definitive terms and get away with it. You are taken as a player. You can assert. You have the authority to frame issues.'

But when she started to live as a woman, all that began to change. Her pay, which had been above average for tenured professors, gradually slipped down to the bottom 10 per cent. She lost her seat on the prestigious University Senate Committee, she found it hard to win grant funding for her research, and she was attacked personally, something that had never happened to her when she was living as a man.

Of course, you might claim that Roughgarden's experience could be put down to transgender discrimination rather than sexism. And it's certainly true that trans women often find it harder to 'pass' than trans men do and suffer discrimination as a result. But, at the same time, Barres was being treated much better, even by those who knew that he was trans.

Roughgarden and Barres used to meet for lunch and compare notes. From her garden in Hawaii, with parakeets squawking in the background, she described to me their parallel but opposite stories. 'We were going through mirror-image experiences. He was beginning to enjoy male privilege and recognizing how much of it he had now acquired, and I was experiencing the reduced influence that I was beginning to have living as a woman.'[7]

'Ben was puzzled and even somewhat offended that his work as a woman wasn't highly regarded; and yet "the same damn work", as he would put it, as a man *was* highly regarded. He was kind of amazed at it, kind of annoyed by it. His career really took off once he transitioned. So he moved toward the centre and my own work moved toward the periphery and we could see ourselves moving in opposite directions.'

What Roughgarden most resented was the decline in respect for her work. 'When I would write an NSF [National Science Foundation] proposal, living as a man, there would be a certain deference to it. It might or might not be funded, but the reviews would be respectful. After I transitioned, the anonymous reviews, both for manuscripts and grant proposals, often had a nasty personal character to them. So academia is a really discouraging environment, I think, for women.

'When living as a male, your views *are* mainstream, you're definitionally mainstream. As a woman, you're not automatically in the mainstream. Ben and I talked about this. When he was living as a woman, his views were regarded as exceptional or out of the mainstream. So he transitions and right away all his work becomes mainstream and he excels in his career. And he was a talented and brilliant man. But on the other hand, he couldn't get that recognition when he was living as a woman. And conversely, when I transitioned, my views were increasingly regarded as exceptional and out of the mainstream.'

It was the propensity for male colleagues to attack the woman rather than her argument that Roughgarden found particularly galling. Living as a man, she had challenged scientific theories and, although some fellow biologists disagreed, she was still taken seriously and offered a tenured professorship. As Joan, the reaction was strikingly different. One fellow scientist shouted at her so aggressively that she thought he was going to come over and hit her. Another invaded the stage after she gave a lecture, yelling at her. Her objectors told her that she hadn't read the literature and suggested they were smarter than her, something that had never happened to her when they saw her as a man. It was as if they were determined to discredit her.

Roughgarden also noticed a completely different dynamic at meetings once she started living as a woman. 'You get interrupted, you need to find a male to support what you said, and not get offended if a male says the same thing and claims credit for it, because at least you got the message out. This is what you have to navigate.' She was discovering what it was like to be at the receiving end of the authority gap, of other people's bias against women.

In an interview with the *New York Times*, Roughgarden described how she was much more frequently interrupted, ignored and condescended to by men, particularly those who had not known her before. 'At first I was amused. I thought, "If women are discriminated against, then I'm darn well going to be discriminated against the same way." Well, the thrill of equal treatment has worn off, I can tell you!'[8] Her conclusion? 'Men are assumed to be competent until proven otherwise, whereas a woman is assumed to be incompetent until she proves otherwise.'

This is exactly the assumption that underpins the authority gap, and it is what both Roughgarden and Barres discovered by transitioning to the opposite sex. *His* competence was suddenly taken for granted; indeed, he was feted for being better than his 'sister', which actually meant better than himself. *She* found herself in exactly the opposite position. Having lost her male privilege, suddenly everything she said was questioned and her competence was challenged.

Sure, again, this evidence is anecdotal. It is an account of two individual life stories. In another way, though, it's almost as scientific as you can get. In each case, every variable has remained the same except one: gender. The critical variable has been isolated. And, in any case, it is supported by much wider studies of the experience of transgender people.

A colleague congratulated his boss for firing Susan, a lawyer, because she was incompetent, and hiring the more skilled 'new guy', who was 'just delightful'.[9] The punchline is that Susan and the new guy were the same person. This is just one example uncovered by Miriam Abelson, a sociologist at Portland State University, who interviewed sixty-six female-to-male trans men in America.[10] Most said they were now seen as more competent, were taken more seriously and had their authority questioned less than before they transitioned.

Abelson concludes: 'A majority of the people I interviewed felt that they had some kind of moment where, if they didn't already believe that sexism existed, this gave them proof.'

Similar research into trans men, by Kristen Schilt from Chicago University, found that the pay of trans women fell by nearly a third after they transitioned but the pay of trans men went up.

One of her interviewees told her that when he expresses an opinion now that he's living as a man, everyone in a meeting writes it down. Another said, 'When I was a woman, no matter how many facts I had, people were like, "Are you sure about that?" It's so strange not to have to defend your positions.' According to a third: 'I used to be considered aggressive. Now I'm considered "take charge". People say, "I love your take-charge attitude."'

Schilt writes, 'Many of the respondents note that they can see clearly, once they become "just one of the guys," that men succeed in the workplace at higher rates than women because of gender stereotypes that privilege masculinity, not because they have greater skill or ability.'[11]

Preston, a blue-collar worker, says of his experience post-transition: 'I swear they let the guys get away with so much stuff! Lazy-ass bastards get away with so much stuff and the women who are working hard, they just get ignored . . . I am really aware of it. And that is one of the reasons that I feel like I have become much more of a feminist since transition. I am just so aware of the difference that my experience has shown me.' He is now taken seriously, without changing his behaviour at all.

Schilt writes: 'Respondents described situations of being ignored, passed over, purposefully put in harm's way, and assumed to be incompetent when they were working as women. However, these same individuals, as men, find themselves with more authority and with their ideas, abilities, and attributes evaluated more positively in the workforce.'

Charlotte Alter interviewed nearly two dozen trans men and discovered the same phenomenon. 'Many trans men I spoke with said they had no idea how rough women had it at work until they transitioned. As soon as they came out as men, they found their missteps minimized and their successes amplified. Often, they say, their words carried more weight: they seemed to gain authority and professional respect overnight. They also saw confirmation of the sexist attitudes they had long suspected: they recalled hearing female colleagues belittled by male bosses, or female job applicants called names.'[12]

Thomas Page McBee is a trans man and an editor at *Quartz*. He has been really struck by how differently he is treated since transitioning. 'When I do talk, people don't just listen: they lean in ... Pretty remarkable for someone who spent thirty years being tolerated (at best) or shunned (at worst) in work environments. I was regularly interrupted. At meetings, my voice didn't prompt people to pause and listen. And I wasn't ever hired, as I was two years ago, for my "potential". All this is despite the fact that I have only worked in progressive environments, places where I have heard men reflect on internalized sexism and where women occupy prominent leadership roles.

'The first time I spoke up in a meeting in my newly low, quiet voice and noticed that sudden, focused attention, I was so uncomfortable that I found myself unable to finish my sentence. I was in Boston, working with a crew of rowdy journalists, in a body that was sprouting hair and muscle and looked, for once, familiarly male to everyone I encountered. It was the most alien I had ever felt. But the room stayed quiet along with me. It was the order of things: everyone in the room waited, men and women alike, for me to open my mouth. I am asked for my opinion near-daily internally and externally, on matters far beyond the realm of my actual job. All of this positive feedback has helped me to become my best, most productive, most creative, most innovative self.'

We see the mirror image for trans women. Daniela Petruzalek was a sales consultant at Oracle when she decided to transition to female. Living as a man, life at work was great: 'I was used to being asked my opinion in every single important decision, no matter what kind of job I was doing. People really wanted to listen to my ideas . . . Everybody wanted to talk to me, get my feedback on their projects, wanted to know what I did to overcome this or that situation. My projects were highly boosted around the company . . . In my last days as a man I was being scouted by other teams to make a lateral move. Also my boss had promised that a promotion would happen very soon. It seemed that I had everything under control.'

And then, after transitioning? 'I was no longer a person to be asked for opinions, the ways I had previously solved my daily challenges were no longer relevant. When I was a man, I was taught to brag about my feats, to take ownership for my winnings . . . when I started doing that as a woman, I would hear remarks like "She's just too arrogant," or, "She's not as good as she says."

'It's really crazy. Technically I was the exact same person. Even better: I was a person that wasn't spending a lot of energy hiding her true self, so I could focus 100 per cent of my energy into the job. That was my best year as a sales consultant, and yet no one seemed to care about me any more. The lateral move offers were withdrawn. The promotion never came. My ideas didn't seem relevant any more.

'I was really new to this "womanhood" thing, so I didn't understand what was really happening. It took me a while but eventually I understood: it was nothing against me, I was just getting exactly what I've asked for . . . *I was being treated as a woman.*'

Paula Stone Williams, a trans woman, married and had children before she transitioned. 'I am learning a lot about what it means to be a female and I am learning a lot about my former gender,' she says. 'There is no way a well-educated white male can know how much the culture is tilted in his favour. There's no way he can understand it because it's all he's ever known and all he ever will know. I never thought I had privilege, but I did.[13]

'Apparently, since I became a female, I have become stupid. The more you're treated as if you don't know what you're talking about, the more you begin to question whether or not you do in fact know what you're talking about. I now understand a woman's tendency to doubt herself.'

Some trans men even find themselves falling into sexist male habits when they interact with women. Thomas Page McBee, who is also the first trans man to box at Madison Square Gardens, admits: 'At work, disturbing patterns came into focus. I kept a tally of how often I tried to get my points across in meetings. Whom did I talk over more often? Women, at a rate of three to one. Even worse, I saw the many subtle ways that I took men just a little more seriously. I was quicker to respond to their emails, more concerned with their perceptions, and more swayed by their arguments.'[14] It's as if there is something in the air that men breathe that leads them collectively to dismiss women more readily than other men. And by joining the brotherhood, McBee started inhaling the vapours.

This was something that Martin R. Schneider, an editor for the movie-reviewing site *Front Row Central*, discovered by chance. He had never quite believed his colleague Nicole Hallberg when she complained that her clients were more difficult to deal with because she was a woman.[15] Then one day he accidentally signed off on an email using her signature. And he instantly noticed how differently this client was treating him. Here's his account of what happened next.

'The client is just being *impossible*. Rude, dismissive, ignoring my questions. Telling me his methods were the industry standards (they weren't) and I couldn't understand the terms he used (I could).' Schneider reintroduced himself to the client as himself, and all the challenging ceased. 'Immediate improvement. Positive reception, thanking me for suggestions, responds promptly and became a model client. Note: My technique and advice never changed. The only difference was that I had a man's name now.'

So, as an experiment, he decided to switch signatures with Nicole for two weeks. 'I was in hell. Everything I asked or suggested was questioned. Clients I could do in my sleep were condescending.'

Meanwhile, his colleague Nicole had the most productive two weeks of her career. 'I realized the reason she took longer [when working under her own name] is because she had to convince clients to respect her. By the time she could get clients to accept that she knew what she was doing, I could get halfway through a job for another client. For me, this was shocking. For her, she was *used* to it. She just figured it was part of her job.'

The female founders of Witchsy, an online art marketplace, became so annoyed by how men treated them online that they invented a male co-founder, Keith Mann. 'It was like night and day,' said one of the founders, Kate Dwyer. 'It would take me days to get a response, but Keith could not only get a response and a status update, but also be asked if he wanted anything else or if there was anything else that Keith needed help with.'[16]

Because bias is often unconscious, because it's like an invisible current in a river, it's so easy – and tempting – to deny it. We don't want to admit that we're prejudiced. And men don't want to concede that privilege has helped them along, until – as for Schneider or the trans women cited – they see it in practice.

But Ben Barres, like many who have lived in both genders, can tell us for sure how pervasive bias really is. Just because you can't see it doesn't mean it's not there. As he wrote in a powerful article in *Nature*: 'Until a person has experienced career-harming bias, they simply don't believe it exists.'[17]

And what these accounts from people who have lived as both

genders show is that we *do* take women less seriously than men. Trans men see a big improvement in the way that people respect them after transitioning; trans women see the reverse. These trans witnesses offer the most persuasive evidence for the existence of the authority gap, for they are the very same people in both incarnations. All that has changed is other people's perception of their gender.

So we need to believe women when they say they are being judged more harshly and taken less seriously than men. Even the men who've become women say so. And we should believe women when they claim that men are treated more leniently and accorded authority and respect by default. That's what the women who've become men experience.

3

The authority gap in action
If you could just let me fini—

MY CLIENT'S JUST
CHECKING THAT YOU
PRONOUNCED HIS
SENTENCE CORRECTLY,
M'LADY

'I used to sell security systems to businesses and was about
to start a lot of work for an old, posh jeweller's. The owner
spent the first meeting refusing to speak to me directly and
when he called me once and got my answerphone I heard him
saying, "Oh for God's sake, it's that bloody girl again, give
me someone who KNOWS WHAT THEY'RE DOING."'

– Michelle, Mumsnet contributor

THANKFULLY, FOR MOST of my working life, I've been relatively well protected from being at the wrong end of the authority gap by virtue of having a high-profile, senior job on a national newspaper. Once you've been very publicly awarded authority by an organization such as *The Times*, it becomes harder for other people to disrespect you, at least to your face.

But I see the authority gap all around me, and as soon as someone doesn't know my history, I notice it snap wide open again.

I was at a high-level international conference recently, talking to two fellow (male) attendees: a former head of the Foreign Office and a BBC foreign correspondent. They know far more about foreign affairs than I do, but they would acknowledge that I have the edge in UK politics, having spent thirty years as a political columnist, most of it for *The Times*. Up came an Italian journalist, who knew none of us. He completely ignored me and asked the men if he could ask them a question about British politics. 'Could Tony Blair ever make a comeback?' he enquired.

'Not a chance,' I replied, and went on to explain why. He half turned his back on me, refusing even to look at me while I was answering, and asked a follow-up question of the two men.

'Look, I'm the British political columnist here,' I retorted, touching him on the arm so that he had to turn my way. 'I *do* actually know what I'm talking about.'

Why should I have to do that? We were all strangers to him, so why would he assume that the men knew more than me, even though we were equal, invited participants at a conference? Why would he then studiedly refuse to engage with my answer? Apart from being incredibly rude, it didn't even serve his own interests.

At another small conference in London, at which we'd all had the chance to see each other in action during the course of the day, I found myself sitting next to a man at dinner whom I hadn't talked to earlier. He was a banker, only a little older than me. He asked me what I did. By then I led a portfolio life and wasn't sure which of my occupations would be of most interest to him. So I replied, 'Well, I do a number of things. I write a political column in the *Independent*, I chair a think tank, I sit on a couple of commercial boards, I make the odd radio programme, I'm on the Council of Tate Modern and I'm involved with some charities.'

'Wow, you're a busy little girl!' he exclaimed.

I was about fifty, older than the then prime minister.

How to respond? The majority view on Twitter was that I should have either stabbed him with my fork or poured my wine over him. Instead I spluttered, 'I haven't been called a "little girl" since I was about six, and I remember it infuriating me even then!'

This is probably the most egregious example of being (literally) belittled that I can recall, at least in my later working life. But these two examples of the disrespect caused by the authority gap help to illustrate what women are up against when they want to be taken seriously. Sometimes it doesn't matter how good your credentials are; you will still be accorded less authority than men. And almost all working women have had experience of one or more manifestations of the authority gap: whether it's being patronized, underestimated, interrupted, challenged unnecessarily, talked over or ignored.

Julia Gillard is a tough, no-nonsense, clearly highly capable woman. You could easily imagine her commanding troops in combat. Yet when she was Prime Minister of Australia she was often patronized by men in a manner that she describes as benign sexism. 'I'd go to any number of meetings of company boards and meetings about Australia's national security, where I was the only woman in the room,' she told me. 'And there would sometimes be men – and they meant it in a very kindly way – whose whole dialogue with you would be, "It must be so *hard* for you, it must be just so *tough* being Prime Minister." And you wanted to say, "Well, look, being Prime Minister's a tough job for anyone. But really, I'm not here to talk about it being a tough job. I'm here to share what I know and to hear what you know. And then hopefully together we can build some solutions to what are complex problems." So there was this assumption that the correct relationship with a woman even who had the power of being Prime Minister was almost nice uncle to favourite niece, rather than a relationship that recognizes that in that room in that moment you were there in your own right.'[1]

Indeed, she was senior to them, I pointed out. 'Yes, exactly!' she replied, laughing.

Michelle Bachelet, now United Nations High Commissioner for Human Rights, has been President of Chile twice. She was tortured and her father was killed by the Pinochet regime, yet she was still brave enough to enter public life herself. Even so, she told me how hard some of her male colleagues found it to accept her authority. 'When I was President for the first time, there was a minister who was older and had been for a long time in Parliament. We had these night meetings of the political committee to decide things. At the end of the meeting, I would make decisions, and sum up how we were going to implement them. And if I said at the end, "OK, we're going to do a, b, and c," he would always take the floor afterwards to say something because *he* had to finish the meeting, to give the impression that *he* was the one who was making the decisions. I could see that, for him, it was a real struggle that a woman was giving the instructions.'[2]

Amber Rudd held three Cabinet jobs under David Cameron and Theresa May, and might have become Chancellor had she not disagreed with Boris Johnson over Brexit. She would be described as a 'political heavyweight' if such a term were ever used of a woman. But that didn't prevent her from being patronized, as she told me: 'I'll never forget the first time I was sitting in the House of Commons, a rather nice elderly Conservative MP, realizing my feminist credentials, turned to me and asked me whether I knew that George Eliot was indeed a woman. And he did it to make friends, but bless them, they're a little way off.'[3]

You don't have to be a senior politician to be *manderestimated*, as I call it. It's hard to find a woman who hasn't experienced it in any walk of life. Polly Marshall Taplin owns and runs a music production company. In 2018, she had sole responsibility for running the second biggest stage at the Glastonbury Festival, that year featuring luminaries such as Stormzy and Liam Gallagher. 'Come Monday morning,' she told me, 'after a draining week's work staying in a caravan onsite, working all the hours God sends, I moved heaven and earth, took taxi and train to Oxford to join a literature summer school in a college there, literature being my first love. What did the literary types say to me? "Oh, did you enjoy being a volunteer at Glastonbury?" The assumption being I had been in first aid or litter-picking. This still stings. Why underestimate me? Just because I am a 58-year-old woman?'[4]

Underestimation of women is particularly common in the traditionally male fields of science, engineering and tech. Silicon Valley is full of geeky guys, some of whom find it hard to believe that women are capable of understanding the complexities of computing. As we saw earlier, many fewer women than men choose to take computer science courses at university, because of the sexism they think they will encounter. For those who brave it, they often find the climate in the workplace afterwards distinctly chilly. They are made to feel as if they are not good enough and as if they don't belong in the 'bro culture'.[5] I met Meredith Broussard, now a data journalism professor at NYU, at a conference at Facebook's headquarters in Menlo Park,

California. She left her job as a computer scientist because of what she calls 'techno-chauvinism'. It infuriated her.

'I couldn't handle the sexism,' she told me. 'All the things that they say about women being edged out of STEM industries are all true and they all happened to me. I blamed myself for it for a long time. I thought, "I'm not strong enough" or "I'm not smart enough" or "I'm not good enough." And then I realized that no, actually, I'm fine but that these social forces are really, really strong.

'People don't take women seriously in tech, particularly if you're a well-dressed, attractive young woman. And that drove me crazy, because *I* knew I knew what I was talking about.'[6] Perhaps you weren't taken seriously because you were young, not because you were a woman? I asked her. No, she countered, 'Young men get taken seriously as technologists. In fact, they get taken more seriously as technologists because there is this cult of genius and this mythology of youth inside tech. This idea of the superiority of the technical solution is tied very closely to the idea of the superiority of the mathematician. And who do we see as a mathematician? A young white guy with a hoodie and jeans, on his laptop with big headphones, changing the world.'

This techno-chauvinism happens even in the highest echelons of tech companies. Shubhi Rao used to be Treasurer of Alphabet, the parent company of Google. She was one of only four women studying computer science engineering at university in the late 1980s, then she spent a pretty torrid time as a systems engineer in a hostile male workplace before doing an MBA.

But things didn't improve much even when she got to the top. 'You have to constantly prove yourself over and over again,' she told me.[7] 'When a guy will say something at a meeting, they're like, "Yes, that sounds great, I think we should do it," and a woman says something and they're like, "Well, I think we should look into it, I think we need more data, we need to think about that, I don't know if that's really going to work." This constant questioning of a woman's expertise is both exhausting and undermines her confidence. Men don't notice its absence – why would they? – so they can enjoy the great privilege of other people accepting their judgements and assertions at face value.

'I get challenged a lot, constantly, and it's become a way of life,' says Rao. 'I know that they're not just going to say, "Because Shubhi said this, with her years of experience and wisdom and expertise, we should agree and go with it." They will challenge and I know they're going to challenge and I'm going to have to be ready for it.'

This reluctance to accord respect to women's credentials is very common, as Helle Thorning-Schmidt has noticed. 'It is very interesting that even when a woman gets a top job that she is qualified for, people ask her what her qualifications are in a way that men are never asked. And I always say we'll have true equality when mediocre women can be promoted to top jobs like mediocre men have been over thousands of years.'[8] How true that is.

Rebecca Solnit, whose essay *Men Explain Things to Me* led to the coining of the phrase 'mansplaining', explains how this disproportionate tendency to challenge women's expertise chips away at their confidence. 'Every woman knows what I'm talking about. It's the presumption that makes it hard, at times, for any woman in any field; that keeps women from speaking up and from being heard when they dare; that crushes young women into silence by indicating, the way harassment on the street does, that this is not their world. It trains us in self-doubt and self-limitation just as it exercises men's unsupported overconfidence.'[9]

In researching this book, I became something of an expert on the subject: an authority, if you like. Yet when I told men what I was writing about I came up against three typical reactions. A small – and very welcome – minority showed interest in the idea and asked me intelligent questions about it, just as I would to someone else writing a book. The vast majority, though, either told me that my basic premise was wrong or else lectured me about the subject, often from positions of complete ignorance. Both reactions were intended to undermine – or *mandermine*, perhaps – my expertise and my authority. Once, a man in the delightful first category was watching while this happened, and he expressed astonishment afterwards, not just at the boorishness of the behaviour, but at the irony of it. Did the mansplainer not realize that he was acting out exactly the patterns of behaviour that *The Authority Gap* was describing? And why did I put up with it? I replied wearily that I was so used to this behaviour that I had learned to nod along, while inwardly hyperventilating with irritation.

For younger women, it can be particularly bad. Laura Bates is an author and founder of the Everyday Sexism Project, a global database of individual women's experiences of sexism and harassment. She may be young(ish), but she is an acknowledged authority in her field. 'For me,' she told me, 'it's a combination of my gender, my age, and my subject, because they are a kind of holy trinity of things that men tend to dismiss: women, young women, and women talking about sexism. It really is quite extraordinary, because it often comes in the context of your being in a place because you are an expert.'[10] And, at thirty-four, she is of an age at which her male contemporaries are rarely patronized.

Bates was once invited to give a presentation to a group of MPs who were meeting specifically to deal with gender equality, and therefore, you might assume would all take the issue seriously. 'I'd been invited to give evidence to them because of the fact that I had curated the largest dataset of its kind that had ever existed of women's experiences of gender inequality. I spoke at length about various forms of harassment, of abuse and sexual violence. And at the end of this meeting, an MP in a very important position came up to me and quietly told me that he felt that I was very "glass half empty", that I had a very negative approach. And that if I really wanted to make change, I should think carefully about ways to make my message more appealing to men. And he felt I sounded quite haranguing.'[11] (I should add that she speaks very calmly and reasonably.)

'And I thought it was really extraordinary because, of course, there's a place to be positive and to celebrate the strides that we've made, but this wasn't that place. This was a place where I had been invited to speak to MPs about the problems women face. There was certainly this sense of an authority gap. Here was an older man telling me, "Don't make a fuss, little girl, you don't really know what you're talking about." And here was I, having been invited to present him hundreds of thousands of women's stories, in their own words, about the reality of their daily lives, which he couldn't know about as a very powerful man.' So, even though this MP was at the meeting specifically to learn about the problems besetting women, and particularly younger women, he couldn't prevent his preconceptions about attractive, youngish, blonde females from sabotaging his appreciation of this aspect of social policy. His resistance to according Bates authority got in the way of his understanding.

Some men never behave as if there is an authority gap between them and the women they encounter. I have had some exemplary male colleagues over the years who have treated me with absolute equality. I have very much enjoyed working with them and I actively notice and appreciate their respect. But the fact that some men behave very well doesn't diminish the problems caused by the people who don't (and that includes some women too).

The older and more senior I get, the easier it is to command authority. And for women all over the world, the authority gap is definitely narrowing, particularly since the #MeToo movement opened people's eyes to the systemic sexism women have had to put up with in their working lives. But it has by no means disappeared. There are still some people who genuinely believe that women are lesser beings, and there are many others whose bias may be unconscious but

nonetheless leads them to undervalue and disrespect female colleagues. They may not notice that they are doing it, but the women at
the receiving end certainly do. And unconscious bias is harder to call
out, as the perpetrator – affronted – is likely to deny its very existence.

Sometimes women get so used to this treatment that they
become inured to it. Major General Sharon Nesmith is the most senior woman in the British army. I visited her in the army headquarters outside Andover, where she sat at her desk in a sandy-coloured
camouflage uniform, with male and female junior soldiers bustling
around her. She is smaller than most of them, but there's no doubt
who's in charge. Her skin has grown so thick that she barely notices
instances of the authority gap, she told me, even though it still
happens at her level.[12] 'There are occasions when I've been to a meeting and people I work with, who are junior to me, have said afterwards they've been surprised about how people have spoken to me or
talked over me or haven't listened to me. And I didn't recognize it in
the meeting. I need people to point it out to me sometimes because I
have become normalized to that behaviour.'

The head of the entire armed forces, Chief of the General Staff
Sir Nick Carter, admitted to me that this is a problem. 'You get a
male-dominated environment where women aren't necessarily
made to feel that they can contribute. So you'll see a committee
meeting which is inevitably going to be 90 per cent male, 80 per
cent if you're lucky, and the extent to which the woman is given her
opportunity to contribute in the same way that the men are needs
working at. And not all men are naturally inclined to do that.'[13]

For some men in the armed forces, women still feel like inter-
lopers who have to be kept at bay – which means that their contributions will be undervalued and often ignored.

'Where you begin to notice it particularly,' he went on, 'is when
you're working at the level I'm working at now, at the Ministry
of Defence, where you've got a lot of female civil servants of signifi-

cant quality. And you're called out pretty soon if you talk over them, or if you don't give them the opportunity to contribute in the way that their male peer might be.' Has that been a culture shock? I asked him. 'Definitely! But on the other hand, this is good. So much of this is about having uncomfortable conversations so you can see the other person's perspective.' He now has two female 'reverse mentors', younger and more junior than he is, who often alert him if he interrupts women or uses language that he might regret. Has it changed his language or behaviour? Yes, he says it has made him more thoughtful.

Major General Nesmith has learned to live with it. In such a male-dominated environment, she hasn't had much choice. Other women, though, find it much more enraging. Bin is a senior British economist. She told me this story: 'I was once talked over on a subject I was the UK expert on by two very senior male journalists. I had been working very closely with the then Chancellor of the Exchequer on this issue and it was the policy topic of the day. They launched into a debate on the topic – it was like watching two deer clash antlers – and I stood there waiting for them to ask me what was going on since they both knew my background. I watched in amazement as they just pointlessly tussled each other with no relevant knowledge. It was the fact that they were journalists that shocked me more than anything. More interested in beating each other than in finding out the truth from the person standing right beside them who was at the event because of their knowledge.'[14]

If it's bad for white women, it's even worse for black ones. Amanda Sesko and Monica Biernat conducted an experiment to see whether people found it harder to recall what black women said after listening to a conversation.[15] Sure enough, they made more mistakes attributing the remarks made by black women than those made by white women, black men or white men. Even more than white women's then, black women's voices go unheard.

This is something that Bernardine Evaristo has often noticed at university meetings: 'They're just hardwired to not want to hear what we have to say and to not want to listen to us,' she told me.[16]

And this seems to be as true of younger men as of their elders. It's not just a generational problem. Shivani, 20, told me how, while working for an alternative investment society at her university, she was made co-director of marketing along with another young woman. The three male executives of the society came to a meeting to discuss marketing. She and her female co-director couldn't get more than a couple of words in, despite being the experts. Almost the entire meeting was spent with the three young men discussing a subject that they knew very little about.[17]

At a trivial level, I often find that my husband insists on independently checking a fact that I know to be true. And I'm not alone. 'I have noticed throughout my life that when I state a fact about something, men will double check it on Google before believing me. I notice they don't do it with men. When men make statements, I find they are automatically assumed to be correct,' says a Mumsnet contributor.[18]

There is also the phenomenon that Joan Roughgarden noticed, when she started living as a woman, that hostile men would often take issue not with her argument but with her as a person. That's something that Liz Truss, Britain's International Trade Secretary, suffers a lot – and it really annoys her.

'I just want people to listen to what I've got to say and argue with the merits of it,' she told me.[19] 'That is one thing that I've found very frustrating. People are often dismissive and will say, "Oh, she's stupid," or, "She doesn't know what she's talking about," rather than arguing against the point I'm making. If you don't agree with me about what the right level of tax should be or how we should reform the education system, fine, argue against that, don't just try and undermine my credibility. It's fundamentally saying, "How dare you even think that you've got the right to say this?" And, my reaction to that is, I'm going to say it even more loudly, because that just pisses me off.'

Mary McAleese also finds these *ad feminam* attacks infuriating. She particularly attracts them when she argues for greater equality in the Catholic Church. 'When I hear theologians saying, "It's not what she says, it's the way she says it," I recognize that as sheer cowardice, because it's a way of deflecting. They respond to me personally, try to diminish me personally, not responding to what I say, and I think that's just such a horrible act of cowardice.'[20]

Having your authority mandermined in all the ways discussed so far in this chapter – being underestimated, ignored, patronized, unnecessarily challenged, or personally attacked – are annoying enough manifestations of the authority gap. What can be more irritating still, though, is being interrupted before you have even had a chance to put your point across. Interruptions are doubly disrespectful: first, they suggest that the person talking over you assumes their views are more interesting than yours, and second, they are a blatant attempt to silence you. The research shows that women are much more likely to be interrupted than men, most of the time by men.[21] Women interrupt too, but their interruptions are more likely to be supportive, agreeing with and amplifying what the other person is saying. They are not intended to shut down their interlocutor. Men are more likely to prevent the other person finishing their point, particularly if she is a woman. And, in general, as we'll see in Chapter 6, men have a tendency to hog the floor, at the expense of women.

And this isn't just a function of their being more senior. It's hard to get more senior and authoritative than being a US Supreme Court Justice. Yet a 2017 study found that, although women made up a third of the justices, they had to put up with two thirds of all interruptions.[22] In other words, they were four times more likely to be interrupted than their male colleagues, 96 per cent of the time by male advocates or justices.

A similar study of the Australian High Court found that female justices were interrupted twice as often as male judges by male advocates, and the advocates interrupted them even more when there was a female Chief Justice presiding.[23] 'The fact that women are more likely to be treated unequally even at the pinnacle of their legal careers suggests an embedded bias towards male judicial authority,' writes the author, Amelia Loughland. 'My qualitative analysis revealed that female judges are subject to the same dominating behaviours in oral argument that they could expect in everyday conversation with men.'

There's a pattern here: of lower-status men interrupting higher-status women. A great example comes from Louise Richardson, vice-chancellor (CEO) of the University of Oxford, the first woman to hold that role in its 800-year history. Sitting in the cavernous eighteenth-century Clarendon Building, with double-height ceilings and enormous windows looking out over medieval college frontages, she looks as if she is queen of all she surveys. But only the previous week, she told me, 'I presided at Congregation and I sit on a throne, chairing a meeting with 350 people, and there were a couple of people there who had a role in it. It was their first time, and I've done it a dozen times. When I was in mid-speech in front of 350 people, this guy said, "You've got it wrong. You should be reading this," pointing to another part of my script, in front of everybody. I said, "Thank you, but actually I've got it right," and I proceeded.

'I said to him the next day, "I'd just like you to give a little thought to one question. Would you have done that yesterday if I were a male vice-chancellor? Would you have interrupted a male VC speaking publicly to a room full of people in a very formal setting to correct what they were saying? Especially when this was your first time at the gathering and, as it turned out, you were wrong?" He was extremely shocked. I said, "I did that mental experiment myself last night. I would just like to ask you to do that mental experiment."'[24]

This is the mental experiment that men should always do when they find themselves interrupting a woman, particularly one who is more senior than they are. For this seems to be a common pattern. There is also research showing that male patients interrupt female doctors, male subordinates interrupt their female bosses and male students their female teachers.[25] It all fits into the patriarchal principle set out by the philosopher Kate Manne in *Down Girl*: that men feel entitled to take what they want from women, in this case the floor, and correspondingly expect that women should cede it to them.[26] This would explain why men often feel entitled to shut down a woman when she is talking, while seeing it as illegitimate or hostile if she interrupts them.[27] For if she interrupts a man, she is taking conversational time, rather than giving it. And if the man is more junior, his interruption is a way of redressing a power imbalance that makes him feel uncomfortable.

Men behave worse when they outnumber women. Put a woman alone in a meeting with four men, and 70 per cent of the interruptions she receives from men are negative.[28] Turn it round so that you have four women and one man in the room: here, just 20 per cent of the interruptions women receive from men are negative. As the study says, when women predominate, 'Men undergo a drastic change. They become far less aggressive.'

Even very small boys – aged between three and a half and five – interrupt girls twice as often as the little girls interrupt them.[29] Parents also interrupt their daughters more than their sons, creating a pattern of behaviour in their children: the boys see that it is legitimate to interrupt girls and the girls learn to expect it.[30] This is setting children the example of the authority gap at such an impressionable age. It is iniquitous, but parents probably don't even realize they are doing it.

You can see how it feeds through into our behaviour in adult life. Professor Mamokgethi Phakeng is vice-chancellor of the University of Cape Town, the most prestigious in Africa. These days she is passionate, confident and fiery, yet she told me how she had to overcome her training in politeness to get herself heard as a postgraduate student, the only female in her maths class. 'I spent six months in class not saying anything because I was raising my hand. The rest of the class, these white kids, were just speaking without raising hands and, after six months, I realized I've got to lose my manners because, otherwise, nobody listens. I learned: they pause, you speak. Otherwise your voice will drown.'[31] This is such a common problem for women, who are trained from childhood to be more agreeable and compliant than boys and are punished if they don't acquiesce. 'Don't chip in!' was a frequent admonition from my parents to me when I, the youngest of four very talkative children, tried to get a word in edgeways at argumentative family meals.

Madeleine Albright, former US Secretary of State, told me that women should learn to interrupt more. 'Because if you raise your hand in a meeting, you usually don't get called on until your point is no longer germane. You need to have what I call "active listening", so you are going to interrupt, you know what you're talking about, and you say it in a strong voice. Then you *must* be a part of that meeting.'[32] Otherwise, it's all too easy to be sidelined.

Kieran Snyder, co-founder and CEO of a tech company, conducted a survey of interruptions in tech company meetings and her findings echo these tips.[33] The men interrupted twice as often as the women did and were almost three times as likely to interrupt a woman as another man. But all the interruptions of a man by a woman came from the three most senior women, and these women were three of the four biggest interrupters in the study. Senior women can only shrink the authority gap in the eyes of their colleagues if they play by male rules.

But even this can be dangerous for them. If women start interrupting more, people often don't like it. This is something Helle

Thorning-Schmidt has discovered: 'Interruption is a big problem for women because you have to be very careful how much you interrupt. And it's so interesting because it's not the same for men. Men can interrupt much more.'[34] They don't suffer the same negative consequences.

Deborah Tannen, professor of linguistics at Georgetown University, has been writing about conversational patterns between men and women since the 1980s. Despite the huge advance of women at work, she is surprised to find that not much has changed. 'I have been giving talks to various organizations, corporations, companies, pretty much non-stop since back then,' she says.[35] 'And whenever I do give these talks, I get the same response. *That's exactly what's happening to me. I experienced that just yesterday. You've just told the story of my life.*

'When I did this research back in the early nineties, I was quite convinced that when there were more women in the workplace, the standards would change. So, in a way, I'm disappointed and also surprised that they haven't. The explanation I would surmise is that a sense of how a person in authority should speak or behave is still based on an image of a man in authority. We still associate authority with men.'

Her advice if a man talks over a woman? 'Don't stop – just keep going.' This can lead to exquisite embarrassment, though, if the man carries on too, and you end up speaking in concert with him for several sentences. Mary Beard, the Cambridge classics professor and TV historian, has her own technique, she told me. 'I say, "Hang on, let me speak. Let the girl speak." I say, "Boys, you've had your turn, let me say something." It's easier as I get older. I do it in a feebly witty way.'[36]

It really helps if other people at a meeting call out this behaviour too, particularly if they are men. If a colleague says, 'Actually, I really want to hear what Rachel is saying,' that can shut the interrupter up. The chair of a meeting needs to be particularly alert to the problem and, if it's pervasive, can instigate a 'no-interruption' rule.

People can also use the Woman Interrupted app, which detects when a female voice is spoken over by a man. According to its findings, this happens 1.67 times a minute in the UK, 1.43 times a minute in the US, and a whopping 8.28 times in Pakistan, 7.22 times in Nigeria and 6.66 times in Malaysia.[37]

There are also ways of ensuring that women's views aren't ignored at meetings. In the early days of the Obama administration, two thirds of the president's top aides were men. The women found that at the daily morning meetings the men simply weren't listening to them. So they adopted what they called their 'amplification strategy': when one woman made a good point, another would repeat it, giving credit to her.[38] This made the men in the room pay more attention, and prevented them taking credit later for the same idea.

When she was Chair of the International Monetary Fund, Christine Lagarde told a World Economic Forum panel, 'When a board member who happens to be a woman takes the floor, guess what? Many of the male board members start to withdraw physically, they start to look at their papers, to look at the floor . . . and you need to disrupt that.'[39] She didn't hesitate to call them out on it: 'When you're the chair, you say, "Somebody's talking. You should be listening."'

It's just as bad in the law. Helen Mountfield is a distinguished QC who is also now principal of Mansfield College, Oxford. She believes the authority gap is embedded in her profession. 'I've become more politically aware of this assumption that what women have to say isn't as interesting,' she told me.[40] 'What happens in court is, there's the interrupting more quickly, the looking bored, the not making notes, and there are lots of little signs that you're not really very interesting or clever, which are really different from the signs that men are given in court.'

And not being listened to can have serious implications for

women's mental health. Anita Martin is a psychiatrist in the north of England, covering a population of 120,000, and she told me, 'It comes up again and again for women when they're depressed that they find it really difficult to be heard, really difficult to assert themselves. Everyone needs agency in their life. If your learned experience is that you're not listened to, that your opinions don't matter, you think there's no point my doing anything because nothing will change. That's the message that they've constantly had from society, that what they think doesn't matter. We see that more in women than men.'[41]

All this behaviour – the authority gap in action – happens because we have a tendency to underestimate women. If we think that they are less capable or expert than they actually are, we are going to pay less attention to what they say. Men in particular tend to discount a woman's views much more than a man's. They manderestimate us routinely.

This underestimation is so commonplace that most women bat it away as if it were a fly buzzing around their head. But it's just as irritating. When I was younger, and an assistant editor at *The Times*, I can't tell you how often I would have the following conversation with a man, the first time we met.

He: What do you do? [*If I was lucky; lots didn't even ask.*]
Me: I'm a journalist.
He: Freelance? [*Start at the bottom . . .*]
Me: No, I'm on the staff.
He: Oh! Who do you write for?
Me: *The Times.*
He: [*voice rises in pitch*] Oh! And what do you do there?
Me: I'm an assistant editor and political columnist.
He: [*even higher pitch*] Oh!

He probably thought he was being flattering, but what it proved was that his initial assumption was that I would be junior and boring, and I surprised him by being neither. As I got older and more confident, I would occasionally ask, 'Just out of interest, if I'd been a man, would you have started by asking me if I was freelance?' Usually, the man would look sheepish and admit he wouldn't.

Of course, there *are* more senior men than senior women in newspaper journalism, and probably more female than male freelancers. And it's also true that some of the best journalists in the world are freelance, though the majority are struggling to make enough to live on. But it is still gratuitously rude to assume that the person standing in front of you is one of those on the bottom rung and, moreover, to say it out loud to her.

Dame Elizabeth Corley is the consummate self-assured City professional. Elegant and *soignée*, with a fearsome brain, she exudes authority. But the former CEO of Allianz Global Investors, described by BAE Systems – on whose board she sits – as 'a globally respected leader in business and finance', told me that she has come up against this too. 'I remember I was commuting to London every day with the finance director of a bank and we'd sit on the train and chat. He wasn't very much interested in what I did. There was a black-tie do at the Guildhall one evening and I was sitting at his table and his face was fantastic! I just loved that moment.'[42]

Most of the highly authoritative women I've interviewed have experienced being mistaken for a secretary or a junior member of a team when they were in fact in charge. Andrea Jung was the first female CEO of the huge American company Avon. She told me that, 'Even when I was the CEO, I had experiences where people were looking around for somebody else because they just assumed I was not the leader of the meeting or the leader of the organization. I think there was just a preconceived notion of who would be the boss, and I didn't fit that mould.'[43] We are so programmed, as the saying goes, to 'Think manager, think male.' And actually, given that Jung is Asian-American, 'Think white male.'

When Anne Mulcahy was running Xerox Corporation, she was one of five CEOs appointed by President George W. Bush to raise money for the victims of devastating earthquakes in Pakistan. Afterwards she was invited to a White House reception as a thank-you. She had to fill in an index card with her name, title and company so that the president would know whom he was meeting. 'I actually joked about it with my husband in line, and said, "You look more like the CEO than me. Let's see what happens,"' she told me. 'Sure enough, I got up to the head of the line and was greeted by the president, and he turned to my husband, whose name is Joe, and he said, "Joe, you're doing an amazing job at Xerox." At which point my husband graciously said, "Well thank you, but that would be my wife."'[44]

Most professional women have the experience of being assumed to be more junior than they are. 'When I turned up at the European Parliament [as an MEP],' Helle Thorning-Schmidt told me, 'everyone mistook me for the secretary. There's nothing wrong with being the secretary, but she's the person who helps other people. She's not the person who hires all the people, the one who takes decisions.'[45]

Often people's racist stereotypes compound their sexist ones, leading to the most terrible behaviour. When Dawn Butler, a black Labour MP, first entered the House of Commons, she walked into a lift and was told by a fellow MP, 'This lift really isn't for cleaners.'[46] To assume that she was a manual worker, uneducated and therefore not qualified to represent her constituents in the nation's legislature, just because of the colour of her skin and her gender, shows how far we still have to go as a society. And to act on that wrong-headed assumption by actively challenging her right to be in a lift shows how pervasively a sense of white male superiority ran through this man's head.

Other interviewees found that, even when people knew who

they were, they were manderestimated. Wan Ling Martello has had a stellar career in business, with senior executive jobs at Nestlé and Walmart, and seats on the boards of Uber and Alibaba. *Forbes* has ranked her ninth in its 'Most Powerful Women International' list. Yet she tells the story of talking to a man she planned to hire while they were walking through Hong Kong airport. 'He stopped dead in his tracks, and I said, "Eddie, what's going on?" He's, like, "Are you really the global CFO and now the CEO of this massive region?" and I looked at Eddie and said, "Yes," and I said, "Why do you say that?" He said, "Because you're so unassuming and you're not what I expect of a senior person at that level." '[47]

Martello believes that showing a certain humility and listening to colleagues, even more junior ones, is a good way of leading and getting better results from subordinates. It's exactly what a senior person *should* be like. As she says, 'The very traits that get us to be underestimated are the very traits that inspire our team.'

So, however much lip-service we pay to gender equality these days, the authority gap still looms large. Its covert bias can be just as damaging, if not more so, than the old-fashioned overt kind. These more subtle acts of discrimination are much more frequent and their effects accumulate quickly over time. Being interrupted, ignored, challenged, talked over, undervalued and underestimated . . . each could be called a micro-aggression, but the macro-effect on women is cumulatively as large as the traditional forms of bias, when women simply weren't allowed to do certain jobs, says a study on the outcomes of different types of discrimination.[48] Like compound interest, the cumulative effect of the authority gap rolls up over a woman's life to produce, by the end, a gaping difference in opportunity and achievement compared with her male contemporaries.

And this covert bias is harder to put a finger on, which makes it more difficult for women to deal with. If a woman is told by her boss that she is not being given a challenging assignment because 'women can't do that sort of thing', she can rightly complain to HR and blame her male manager for being a dinosaur. If, instead, he says, 'I don't think you're ready to take on this challenge,' she is more likely to blame herself, to lose confidence, and to believe his assessment, even if it is coloured by unconscious bias. In other words, both of them may end up believing that the man's inaccurate judgement is accurate, even if it isn't. And that is how women get unfairly held back.

It doesn't have to be like this. It may be hard for us to change our unconscious bias (though it is gradually decreasing over time as we become more used to women in authority), but we *can* look out for it, recognize it when it tricks our brain, and then correct for it. At the end of the book, I'm going to give you all sorts of ways in which we can each, individually, start to narrow the authority gap, and set out how employers, teachers, the media and the government can help.

In the meantime, though, if you're a man, you might, justifiably, be asking yourself, 'Well, why should I want to narrow the authority gap? What's in it for me?' The surprising answer is: 'A lot.'

4

It's not a zero-sum game
We all gain from narrowing
the authority gap

'Gender equality is also in our interests as men. If you listen to what men say about what they want in their lives, gender equality is actually the way for us to get the lives we want to live.'

– Michael Kimmel, sociologist

F YOU'RE A man reading this book, may I first of all say thank you. You are unusual in picking up a book written by a woman, and even more unusual in being prepared to read a book primarily about women. Your commitment will be rewarded, for this chapter is designed for you. Counterintuitively, perhaps, it is mainly about what you can gain, as a man, from taking women more seriously and treating them with equal respect. There is a huge amount of evidence to suggest that this is a positive-sum game in which everybody wins.

I know that may sound odd. Isn't gender equality a seesaw in which, if one side rises, the other must, by definition, fall? I don't deny that there may be individual instances in which, if you are in direct competition with a woman for a job or a promotion and the bias against her is dissolved, you may find that she beats you on merit. But in pretty much every other aspect of your life, including your day-to-day interactions at home and at work, gender equality is likely to make you happier, healthier and more satisfied. You'll even sleep more soundly *and*, as a clincher, have more and better sex.[1]

Let's start from the individual and work outwards to the workplace, the wider economy, the country as a whole and the planet, to see what we can all gain from narrowing the authority gap.

A man who treats a woman with equal respect in everyday interactions will find that the two of them will get on significantly better, as friends, as colleagues, or even as potential partners. A friend of mine's son came back from university once and exclaimed to his father, 'I've discovered the secret of pulling girls, Dad.' 'What is it?'

asked my friend. 'It's easy,' said Tom. 'All you have to do is listen to what they say!'

It's so true. A woman can tell within just a few minutes of meeting a man whether he is genuinely treating her as an equal or is patronizing her and stops listening as soon as she opens her mouth to plan what he is going to say next about himself. Why would she be interested in the second type, either as a friend or a lover, if he clearly thinks he is superior and has no interest in her as a person? Believe me, being sexist is not a good look for a man.

Women these days are much more inclined to choose as romantic partners men they think will be good and involved fathers and share the second shift equally than the traditional husband who expects the woman to do all the chores and childcare.[2] As the psychotherapist Phillip Hodson puts it in *Men: An investigation into the emotional male*, 'Men only need to change a little to gain great improvements in their relationships but they falsely see this change as considerable and resist it.'[3]

Reams of research shows that everyone in a heterosexual family, including the man himself, gains from a man sharing in a more egalitarian partnership. The woman is more satisfied with the relationship, the couple communicate better, she feels respected and part of a genuine team, and she is less resentful and exhausted by taking a disproportionate share of the unpaid work. As a result, she is happier and healthier. So are their children, who also display fewer behavioural difficulties and do better at school. But, best of all for these purposes, the men themselves are also happier and healthier: they are twice as likely to be satisfied with life, they smoke less, drink less, take fewer drugs, suffer less mental ill health, are less likely to get divorced, have a better relationship with their children – and they get significantly more frequent and better sex.[4] What's not to like?

The good news, for both men and women, is that men are increasingly keen to play a bigger part in their children's lives. Some

70 per cent of fathers say they don't spend as much time as they would like with their young children.[5] Even before the pandemic, nearly as many men as women were looking for jobs in which they could work flexibly.[6] Fathers who do work flexibly are more satisfied with their work–life balance and are less likely to leave their jobs.[7] And their partners are nearly twice as likely to advance in their careers as partners of men who don't work flexibly.[8] Meanwhile, fathers who take parental leave – even short paternity leave – have a much closer relationship with their children.[9]

Nor is it considered unmanly these days to be a good and involved father. Three quarters of people globally *disagree* that a man who stays at home to look after his children is less of a man, compared with just 18 per cent agreeing. In the UK, the figures are 81 per cent to 13 per cent.[10]

Being a good father these days is thought to be just as important as being a breadwinner. In 2013, J. Walter Thompson Intelligence asked 500 British men to identify 'the primary things that define men today'. Among their responses, 'providing financial support for family' (51 per cent) was rated by the men barely ahead of 'parenting abilities' (49 per cent) and 'providing emotional support for family' (46 per cent).[11]

Fathers being more involved with bringing up their children doesn't just free up women to advance at work, thus helping narrow the authority gap. It also transforms attitudes in the next generation. Daughters with dads who do their fair share are more likely to pursue their career aspirations, often in less stereotypical occupations, with more self-esteem and self-confidence. Sons who see their fathers share the household duties equally have a more egalitarian perspective of women's and men's roles at home and work.[12] And when they become teenagers, these boys are half as likely to be violent as their peers who have rigid views about masculinity and gender.[13]

It seems that gender equality suits men rather well. It allows them to experience all the love and comfort that come with stable relationships and happy families. It allows them to escape the rigid constraints of masculinity that came with the old notions of patriarchy, which can be just as unpleasant for men as for women.

As Julia Gillard, former Australian prime minister, put it to me: 'I am sure that there are men who also feel in their own lives the negative impact of gender stereotyping. If they were able to be exactly the person they want to be with no adverse reactions from others, then they might make quite different choices. For example, a man might want to say, "I'd prefer to work part time because I really want more hours with my children," but doesn't because he worries that if he says that out loud in his current workplace he is going to be viewed as not having the ambition it takes to succeed.'[14]

Cherie Booth, the human rights lawyer also known as Mrs Tony Blair, agrees: 'The truth is that the patriarchy has disadvantages for men too. There are a lot of men who actually do want to spend more time with their children or maybe don't like aggressive ways of leadership, who need to be able to feel that they can do that without being designated a second-class, female citizen. The alpha male is a bully, not just to women, but to other men too.'[15]

The Norwegian sociologist Øystein Gullvåg Holter has written a wonderful paper called 'What's in It for Men?', in which he enumerates all the benefits that men win in more gender-equal European countries and more gender-equal US states. They are less likely to get divorced. Their chances of dying a violent death are almost halved. The gap between male and female suicide rates is narrower. Men are also less likely to be violent against their partners and children, which in turn reduces the children's risk of being violent in later life. Best of all, though, they are happier.

'It is a common misunderstanding that increased gender equality provides benefits and privileges for women at the expense of men's benefits and privileges,' he says. In fact, he finds, men in more gender-equal countries and US states are twice as likely to be happy, and nearly half as likely to be depressed. This holds true whatever their class or income.[16]

So if men are happier at home in more gender-equal households, how about at work? Well, my observation above holds just as true in the workplace. If you treat your female colleagues with equal respect and value their competence as highly as that of men, they will like you more, work harder for you and be less likely to leave their jobs. And if you are lucky enough to have a female boss, you will probably find her a better people manager. Gallup's *State of the American Manager* report surveyed 27 million employees and found that those who work for a female manager are 26 per cent more likely than those who work for a man to strongly agree that, 'There is someone at work who encourages my development,' and 29 per cent more likely to strongly agree that, 'In the last six months, someone at work has talked to me about my progress.'[17]

As a result, people who work for women tend to be more engaged and loyal, which is good news for employers too. Both men and women are more engaged with female bosses, but the biggest gap is between women who work for women (35 per cent engaged) and men who work for men (only 25 per cent). 'Overall,' says the report, 'female managers eclipse their male counterparts at setting basic expectations for their employees, building relationships with their subordinates, encouraging a positive team environment and providing employees with opportunities to develop within their careers.'

And the female managers themselves are more engaged than their male counterparts, finds the survey, perhaps because they know they have to work harder to gain the same recognition. This is something that Mike Rann, former premier of South Australia, has noticed in politics. I went to talk to him about the appalling misogyny that his friend and colleague Julia Gillard had endured when she was Australia's first and so far only female prime minister. But he expanded on his theme. 'Women read their briefs, they don't just read the summary of their Cabinet papers, they've actually done the homework, often much more diligently,' he told me.[18] 'And why? Partly because it's the right thing to do, but because they're constantly being judged more harshly, under different standards to the blokes, they have to make sure they go the extra mile. So I think men have a lot to learn from women and I don't understand why they're so scared.

'Women coming in greater numbers into our parliaments and Cabinets has made politicians more professional than perhaps they were. Before, they were enthusiastic amateurs. The men are much more likely these days in the presence of women to conduct themselves better.'

One consequence of the authority gap is that women are held to higher standards. This means that employers are often losing out on under-promoted female talent. So they, too, have a lot to gain from narrowing the gap. As Rann points out, women often outperform men. For instance, houses listed by female estate agents sell for higher prices, female lawyers are less likely to behave unethically, and patients treated by female doctors are less likely to die or be readmitted to hospital.[19]

The business case for promoting more women to positions of authority is very strong. According to McKinsey & Company's 2019 report on the subject, which looked at more than 1,000 large companies in 15 countries, the most gender-diverse companies were 25 per cent more likely to earn above-average profits than the ones with very few women. And the more women there were in senior jobs in a business, the higher the likelihood of outperformance.[20]

Another study of FTSE350 companies found that those in which women make up more than a third of their most senior jobs have a net profit margin over ten times greater than companies with no women at this level.[21] That, as economists say, is non-trivial. Think how much richer the country would be if more women played a bigger role in running businesses. That would mean more jobs and higher wages for us all.

Investors and the stock market understand this. Many big institutional investors are now putting pressure on companies that have very few senior women. This isn't for box-ticking purposes, but to increase shareholder value. Once a company (or any other employer) has more senior women in its ranks, it is likely to be able to recruit better talent. When considering a potential employer, 61 per cent of women look at the diversity of the employer's leadership team and 67 per cent at whether it has positive role models similar to them.[22]

One reason for the outperformance of these gender-diverse companies is that they are fishing in a much larger talent pool, and women often outperform their male colleagues.[23] But there is also strong evidence that more diverse teams (including race, nationality and class as well as gender) make better decisions, even if the members don't always feel it at the time.

Having an outsider come into a team may be uncomfortable to start with, but it is that very discomfort that jolts us from our tramlines. While homogeneous groups may feel more confident that they have made the right decisions, it is the diverse groups that actually perform better. Katherine Phillips, a professor of business and organization, did an experiment putting people into groups investigating a murder. In some of them, the members all knew each other, but the ones that contained outsiders were more likely to find the right suspect because they ended up thinking harder about the problem.[24] The least diverse groups were much more confident about their decisions, even though they were more likely to be wrong. 'Generally speaking, people would prefer to spend time with others who agree with them rather than disagree with them,' says Phillips. But agreeing with each other does not always produce the best results. 'When you think about diversity,' she goes on, 'it often comes with more cognitive processing and more exchange of information and more perceptions of conflict. It's kind of surprising how difficult it is for people to actually see the benefit of the conversations they are having in a diverse setting. When these diverse groups perform well, they don't recognize their improved performance.' But it is there.

Venture capital is a famously cliquey male field. But VC firms that hire more women as partners have 10 per cent more profitable exits. Another study finds that women-run private tech companies earn a 35 per cent higher return on investment.[25] Yet companies run by men still win 93 per cent of all venture capital funding.

So just think how much richer we could all be if we used this potential better. Giving women more authority – taking their talents more seriously, promoting them more, lending them money, allowing them to lead – could hugely boost the world economy.

Christine Lagarde, the economist who is now President of the European Central Bank, has co-authored a paper with Jonathan Ostry which calculates that, on the basis of the complementary skills and perspectives that women bring to the workforce, countries ranked in the bottom 50 per cent for gender inequality could boost their GDP by 35 per cent if they closed the gender gap. What's more, this could actually increase men's wages because having more talented women in the workforce would lead to higher productivity, from which everybody gains.[26]

McKinsey, meanwhile, estimates that, if all countries in a region matched the rate of gender-equality improvement of the best one, this could add $12 trillion, or 11 per cent, to annual global GDP: equivalent to the current GDP of Germany, Japan and the UK combined.[27] There are massive gains to be had.

The world might also be better run if women were accorded as much political authority as men. We have already seen how success-ful leaders such as Jacinda Ardern in New Zealand can be, with their more consensual and empathetic style. But this isn't just anecdotal. Research shows that female politicians, on average, do more con-stituency work than men, are less corrupt and have a leadership style that is more cooperative and inclusive. They focus more on issues that help the most vulnerable. They also win more central govern-ment spending for their constituencies and they legislate more.[28]

And we have more chance of living in peace. Countries that have more women in power are less likely to go to war and less likely to have a civil war. Countries with 10 per cent of women in the labour force are nearly thirty times more likely to experience internal con-flict than countries with 40 per cent. Meanwhile, including women in peace processes makes them more successful and long-lasting.[29]

In the wonderfully named *Journal of Happiness Studies* is an art-icle called '(E)Quality of Life', in which the authors conducted a cross-national analysis of the effect of gender equality on satisfac-tion with life. They write: 'By any standard, improvements in the status of women appear to be associated with large improvements in the overall quality of life within a nation. The conclusion is fairly straightforward: across our different measures of the relative em-powerment of women, the data suggest that society is happier as women achieve greater equality.'[30] No wonder men as well as women prefer living in Sweden than in Saudi Arabia.

'It could be that any improvement in the wellbeing of women produces a corresponding reduction in satisfaction among men, as if quality of life is a zero-sum game in which improvement for some means a diminution for others,' they concede. But no: 'For both men and women, gender equality would seem to lead to greater life satisfaction regardless of the measure used.' Given how wor-ried we are these days about teenagers' wellbeing, it is cheering also to discover that adolescent girls *and boys* are happier in more gender-equal countries, even after controlling for national wealth and income equality.[31]

Finally, let's consider the implications for the future of the planet. Women are more likely to worry about climate change and to believe that it will harm future generations. They are also more likely to believe that it will affect them personally.[32] So having more women in positions of decision-making power, with people listening to them, would help to reduce global warming.

At the local level, an experiment with forest-users in Indonesia, Peru and Tanzania found that including at least 50 per cent women in the decision-making groups led to more trees being conserved and to payments being distributed more equitably.[33] At the national level, too, it makes a difference. Having more women in national parliaments leads to tougher climate-change policies and lower carbon dioxide emissions.[34]

So, from the home to the workplace, the economy, the nation state and the planet, allowing women to have equal authority to men is in *all* our interests. We gain so much, men as well as women, from having the added talent and perspective of women contributing to our shared lives. We will be happier, healthier, richer, more fulfilled and better governed if we close the authority gap. We might even save the planet in the process.

The confidence trick

Confidence is *not* the same as competence

'Because we . . . commonly misinterpret displays of
confidence as a sign of competence, we are fooled into
believing that men are better leaders than women.'

– Tomas Chamorro-Premuzic,
organizational psychologist

HAVING SET OUT what the authority gap is, and how much we could gain from narrowing it, we are now going to look, over the next few chapters, at what lies at its heart – and what we can do about it. If women are as talented as men and have all the right personality traits to be good leaders, maybe there's another reason why we don't rate them as highly or listen as carefully to what they say. They are often not as confident – or maybe, they're not as full of bullshit. Looking back to my discussion in Chapter 3 of the ways in which women's confidence is undermined, it's perhaps not surprising that this is the case.

When Janet Yellen, now US Treasury Secretary, was Chair of the US Federal Reserve, she was the most powerful woman in the world – or perhaps equal first, tied with the German chancellor Angela Merkel. Yellen ran the American economy, which meant, in practice, running most of the world economy. Yet she still felt like an imposter in the job.

'I've certainly questioned, "What am I doing here? How did I get myself in this situation?" ' she admitted to me when I went to visit her in Washington DC. 'I've felt like, "I can't do this." '[1] She isn't alone. Many of the highly successful women I've interviewed for this book have confessed to similar anxieties. Women leaders suffer from impostor syndrome more than twice as often as men, and research suggests that unconscious bias is a big contributor.[2]

Baroness (Brenda) Hale, former President of the UK Supreme Court, the most senior judge in the land, has had this worry at every stage of her life. 'I was never sure whether I was going to be able to do the next thing that I did, so you pass the 11-plus and you go to the high school, am I going to be able to cope in the high school? It turns out you can. I get into Cambridge, am I going to be able to cope in Cambridge? It turns out you can. I go to the bar in Manchester, am I going to be able to cope? It turns out you can. I go to the Law Commission, am I going to be able to cope? It turns out you can. So, it's more me thinking whether I could do it or not and finding out, yes, it did seem to be possible.'[3] It certainly was possible, but think how morale-sapping it must be to spend your whole life plagued like this with self-doubt.

Even Christine Lagarde, one of the most authoritative women in the world, has admitted: 'I would often get nervous about presentations or speaking, and there were moments when I had to screw up my courage to raise my hand or make a point, rather than hanging back.'[4]

Of course, many men experience impostor syndrome too, though they may be less likely to admit to it. When I asked my husband if he ever suffered from it, he said he did, but in a completely different way from how I described it in women. If he knows he is not expert enough in a subject, he sees it as an exciting risk, like taking a corner fractionally too fast. Will he be able to blag successfully enough to pull it off? That is very far from the feeling that you don't deserve to be there in the first place.

The underlying issue is confidence. Impostor syndrome is a symptom of lack of self-confidence. So maybe the authority gap exists because women don't tend to display as much confidence as men? It's certainly true that girls and women are, on the whole, less confident than boys and men. This is perhaps not surprising, as they are treated differently from birth. As small babies, mothers tend to be more protective of their daughters and fathers are more likely to throw their sons up in the air and catch them.[5] Differing treatment such as this helps to give boys more physical confidence.

But it's also true of intellectual confidence, which is vital if we want people to respect our views and see us as authoritative. As we heard earlier, British parents think their sons are cleverer than their daughters.[6] And adult men think they are cleverer than adult women.[7] Yet the distribution of IQs between the genders is identical, except at the extreme ends of the curve.

Alan Ryan used to be Warden of New College, Oxford. So perplexed was he by the discrepancy between women and men getting first-class degrees at Oxford, despite their apparent intellectual similarity, that he decided the college should do some psychometric testing on its students. The results were startling, he says. 'The women were normal and the men were crazy. The men all overestimated how interesting they were, how intelligent they were, how much people liked them. They were skewed at the self-deceiving end of the spectrum, while the women were right on the centre. You are more likely to get boys in a tutorial saying, "Look at me! Look at me!" and getting more attention.'[8]

Even among today's students, who ought to be young enough to know better, there is this underestimation of young women's intelligence and overestimation of young men's. A recent study asked biology students who was the smartest and best-informed in their class.[9] Male students consistently rated other male students as cleverer than better-performing female ones. This male bias increased over the course of the term and persisted even after controlling for class performance and outspokenness. The women, meanwhile, rated other students accurately.

This constant mandermining of women's ability is bound to dent their intellectual confidence. And the effect starts surprisingly early. Lin Bian, a psychologist at the University of Illinois, read out a story to 240 five- to seven-year-old children. 'There are lots of people at the place where I work, but there is one person who is really special. This person is really, really smart. This person figures out how to do things quickly and comes up with answers much faster than anyone else.'[10]

She then showed them photos of two men and two women and asked them to guess which was the 'really, really smart' person. At the age of five, both boys and girls chose an adult of their own gender. But once the girls reached six or seven, they started to assign brilliance more readily to men. The same effect was found when they were asked to guess whether a 'really, really smart' child was a girl or a boy. And this was despite them knowing that girls tended to get higher grades in school.

Then Bian introduced the six- and seven-year-old children to two games: one for 'really, really smart' children and one for 'children who try really, really hard' and asked them whether they'd like to take part. The girls were much less likely than the boys to want to play the 'smart' game, but just as keen to play the 'hard-working' one. When the children were asked to select team-mates from photos of children they didn't know, the odds of them selecting a girl rather than a boy dropped by 51 per cent when the game was said to be for 'really smart' children. So children are absorbing the false stereotype that boys are cleverer than girls from a really early age and internalizing it as true. No wonder girls grow up less confident of their intellectual abilities than boys.

You see the same happening among adults. Bian asked men and women to refer people for one job that required high-level intellectual ability and one that required someone who was well motivated. The odds of referring a woman rather than a man were 38 per cent lower for the job that demanded serious intelligence, and women were just as biased as men.

Both children and adults, male and female, from countries all over the world, also associated 'brilliant' much more with men when they took an Implicit Association Test, which seeks to measure unconscious bias. In fact, 'brilliant' was second only to 'strong' in its association with male. No other psychological trait came near.[11]

And it's probably because of the way parents and teachers perceive children. Here's an example from a 20-year-old (female) American student: 'In high school I had a 4.36 grade point average and was constantly praised for my "hard work", "dedication", "work ethic", and worst of all, "grit". My high-achieving male friends were praised for "genius", "brilliance", and "talent". The catch is, I didn't study, I didn't do the assigned reading, I didn't review the notes, I just showed up to class, and wrote down the answers later. Yet my work was still held up by teachers as an example of what happens when you study hard and take notes and read all the extra textbook chapters. Despite top scores in science, literally not one single adult ever suggested that I consider a STEM career.'

As Janet Yellen – who has a dazzlingly brilliant brain but would never admit it – told me, 'You hear people talking about respecting people who are brilliant, and that tends to be something that people perceive much more to be true of men than women. You rarely hear somebody say that a woman is brilliant; women are hardworking, whatever, but they tend not to be viewed as brilliant. Yet I see no reason to think that there are fewer brilliant women than men.'[12]

Not only does this association of male with brilliance put girls and women off subjects that are thought to need it, such as maths, physics, philosophy and economics, it also means that the few who brave the subjects nonetheless are often written off by men when they get there, however bright they are. And that is bound to chip away at their intellectual confidence.

A recent survey by the American Economics Association found that half the female economists interviewed said they had been treated unfairly because of their sex, compared with only 3 per cent of the men.[13] A startling 70 per cent of the women said their colleagues' work was taken more seriously than their own.

When you discover what male economists think of their female colleagues, this is perhaps less surprising. Alice H. Wu of Berkeley University did a text mining search of more than a million posts on the Economic Job Rumors website.[14] It's the digital equivalent of a water cooler, where young economists gossip about candidates and vacancies. The thirty words most often used about women economists are almost too horrible to print. They are, in order: 'hotter', 'lesbian', 'bb' (internet speak for 'baby'), 'sexism', 'tits', 'anal', 'marrying', 'feminazi', 'slut', 'hot', 'vagina', 'boobs', 'pregnant', 'pregnancy', 'cute', 'marry', 'levy', 'gorgeous', 'horny', 'crush', 'beautiful', 'secretary', 'dump', 'shopping', 'date', 'nonprofit', 'intentions', 'sexy', 'dated' and 'prostitute'.

The ones used about men are mainly to do with economics. There *are* words in there such as 'juicy' and 'bully', but also 'adviser', 'Austrian' (a school of thought in economics) 'mathematician', 'pricing', 'textbook', 'Wharton' (the University of Pennsylvania business school), 'amusing', 'goals', 'greatest' and 'Nobel'.

This is not just disgusting, but genuinely shocking. What were these male commenters thinking of? Well, sex, clearly. But do they really have no respect for, or interest in, the intellectual qualities of their female colleagues? Think how hard it must be for a female economist to thrive in a world of such intellectual rigour when the words most likely to be associated with her are 'tits', 'anal', 'horny' and 'prostitute'. And these are written in a public forum!

Yellen herself was, at the time of writing, President of the American Economics Association, and she was horrified by these findings. But not surprised. In her first academic job at Harvard, she was the only woman in the economics faculty. 'That was, I think, the only real phase in my career where I felt that, wow, being a woman in a man's world is really unconducive to success,' she told me. 'It's very isolating and it's very difficult to succeed in an environment that is that aggressive, where people are that self-confident. They look down on you and assume that you're not going to measure up, and really don't have any interest or sense of respect for you.' None of the men wanted to collaborate with her, and economics is a subject in which collaboration is crucial for career progression. It was only when another woman joined the department and they wrote papers together that her career took off.

Yellen told me stories of how female economists are often treated in seminars. 'Seminars in economics can be very aggressive: people like to show off, they like to show they're smart and exhibit their brilliance. So you come in to make a presentation and you flash up your first slide that says, "Here's what I'm going to do in this seminar." And before you've even laid out what you're planning to do, someone, usually a very aggressive guy, will say, "Well, you asked the wrong question, and the way you're going to go about it is not going to prove anything," even if it was an interesting question that you were looking at.

'They do this disproportionately to women. To a guy, they'll say, "Well, that's an interesting way of looking at things, I wouldn't have looked at it that way myself." Something more encouraging.'

So you can quickly see how a 'guy' in economics will find it much easier to maintain and build confidence than a woman. His male colleagues will take him more seriously and challenge him less. His female counterparts, meanwhile, will find their colleagues paying more attention to their bra size than to their microeconomic theories. And when these women do perform in an intellectual context, they are more likely to be shot down. No wonder they feel less confident.

Philosophy, another academic discipline in which brilliance is prized, can be equally confidence-destroying for women. Sally Haslanger, professor of philosophy at MIT, wrote in an influential article in the feminist philosophy journal *Hypatia*: 'There is a deep well of rage inside of me; rage about how I as an individual have been treated in philosophy; rage about how others I know have been treated; and rage about the conditions that I'm sure affect many women and minorities in philosophy, and caused many others to leave.'[15]

In graduate school, she wrote, 'I was told by one of my teachers that he had "never seen a first-rate woman philosopher and never expected to because women were incapable of having seminal ideas".' This is a senior academic displaying not unconscious but blatantly conscious bias. Even if women do start off as intellectually confident as their male colleagues, they would need granite-like self-assurance to shrug off such assaults on their self-esteem. Yet the chances are that they don't, because of the way they have been taught in childhood.

Girls are expected to be quieter and better behaved in class, and they get less attention and encouragement from teachers. One American study found that elementary and middle school boys had eight times more classroom attention than girls.[16] When boys called out in class, teachers listened and responded, but when girls did, they were told to 'raise your hand if you want to speak'. When boys didn't volunteer answers, the teachers were more likely to urge them, rather than the girls, to give a response or an opinion.

According to David and Myra Sadker, who studied this classroom behaviour, 'Teachers talk less to girls, question them less, praise, probe, clarify and correct them less. Female students accept the leftovers of teacher time and attention, and morsels of amorphous feedback. As a result, most girls learn to mind themselves, stay out of mischief and settle for a quiet role in the classroom. Girls quickly learn to smile, work quietly, be neat, defer to boys and talk only when spoken to . . . Little wonder that so many girls lose their voice, confidence and ambition, a problem likely to haunt them in adulthood.'

Allyson Julé, a professor at Trinity Western University, has conducted a similar study, and found that teachers repeat boys' comments as recognition of their contributions nine times more than girls', address questions much more to boys than to girls, and praise boys more for their answers. She agrees that teachers' behaviour has a dramatic effect on girls, not just at the time but in later life too.[17] 'From my research, the teacher does systematic things that can silence [girls]. If she'd done different things, they would have said more. It's not that girls are quiet, it's that girls have been silenced.'

'What if school is a confidence factory for our sons, but only a competence factory for our daughters?' asks the clinical psychologist Lisa Damour.[18] 'What if those same habits that propel girls to the top of their class – their hyper-conscientiousness about schoolwork – also hold them back in the work force?' She believes that girls rely too much on 'intellectual elbow grease' while boys often get by on blagging and doing as little as possible to keep parents and teachers off their backs. And teachers encourage these behaviours.

As she writes: 'That experience – of succeeding in school while exerting minimal or moderate effort – is a potentially crucial one. It may help our sons develop confidence, as they see how much they can accomplish simply by counting on their wits. For them, school serves as a test track, where they build their belief in their abilities and grow increasingly at ease relying on them.'

If girls do coast along without making much effort at school, though, teachers often hate it. One of my daughters was a consummate blagger, getting by academically almost on wits alone. Her technique succeeded: she always managed to scrape top grades. But I remember one teacher complaining furiously, 'We've never had a girl like this before!' I retorted that they must have had plenty of boys like that. The teacher had to concede that this was true. But somehow it was different, because girls were held to a different standard.

Girls are also taught, by parents, teachers and their own peers, to be more modest and self-deprecating than boys. Watch little boys play and talk together and a lot of it consists of boastful competitiveness: 'My dad's got a bigger car than yours'; 'I can kick the ball further than you.' Girls are more likely to do themselves down – 'I'm hopeless at maths'; 'I hate my hair' – in order to gain approbation from other girls. Female bonding consists of admitting vulnerability to each other; the very opposite of male bonding. And it continues into adulthood.

But the trouble is that if men are modest, we tend to assume that they must be better than they claim; they're just being charmingly self-deprecating. Whereas if women are modest, we take them at their word. You say you're bad at maths? I believe you. As Deborah Cameron, Oxford Professor of Language and Communication, writes: 'Women have often discovered that a symbolic display of humility from them is interpreted less as principled egalitarianism and more as a confirmation of their assumed inferior status.'[19]

Yet if girls and women aren't modest, they are often penalized for lacking humility (by other girls and women as well as men). When I first became opinion editor at *The Times*, I found myself to be one of only two women among about twenty men at morning and afternoon conference, when the next day's paper was planned. If Bridget, the features editor, was away, I was the only one. National newspapers are furiously competitive places of work, and most of my male colleagues were desperate to prove that they knew everything, even if they didn't. And there were no women more senior than me to act as role models. How was I to navigate this maelstrom of masculinity?

I quickly realized I had two choices. I could play the demure female, in which case I would be rolled over by these ultra-competitive and aggressive men. Or I could sit up straight, fight my corner, and appear, at least outwardly, as confident as they were. The latter seemed the only viable option.

And, in some ways, it worked. One of my early experiences there involved my boss standing with his face just inches from mine and yelling at me. I didn't flinch, and he never tried bullying me again. Yet, because I was prepared to act as confidently as my male colleagues, to act as if I had just as much right to be in the room as they did, it's perhaps no coincidence that the satirical magazine *Private Eye* ran a regular column, with an illustration of me at the top, caricaturing me as 'Mary Ann Bighead'. They might as well have subtitled it 'Woman, Know Your Place'.

Boys and men can much more easily get away with being boastful or bombastic. As a result, you're much more likely to see a teenage boy bullshitting than a teenage girl. Lest you think I'm generalizing, take a look at a study of 40,000 fifteen-year-olds in nine countries called, unusually for an academic paper, 'Bullshitters. Who are they and what do we know about their lives?'[20]

These students were given a list of sixteen mathematical concepts and asked to rate their knowledge of each of them, from 'never heard of it' to 'know it well, understand the concept'. Unbeknown to the teenagers, the researchers had inserted three fake concepts– 'proper number', 'subjunctive scaling' and 'declarative fraction' – into the list. In all nine countries, boys were much more likely than girls to claim that they knew and understood the fake concepts.

What's more, the bullshitters believed their own bullshit. 'Our study shows that bullshitters express much higher levels of self-confidence in their skills than non-bullshitters, even when they are of equal academic ability,' said Nikki Shure, co-author of the study.

So this leads to a double bind for women. Either they appear as confident as men, running the risk of being disliked, or they do themselves down. Yet modesty too is damaging, for other people are quick to mistake confidence for competence – and therefore under-confidence for lack of competence. If someone is super-confident about their ability, we tend to believe them, particularly if they are male. The psychologist Tomas Chamurro-Premuzic wrote an article in the *Harvard Business Review* which led to his book *Why Do So Many Incompetent Men Become Leaders?*.[21] His answer? 'In my view, the main reason for the uneven management sex ratio is our inability to discern between confidence and competence. That is, because we (people in general) commonly misinterpret displays of confidence as a sign of competence, we are fooled into believing that men are better leaders than women.'

Yet, as he points out, 'Arrogance and overconfidence are inversely related to leadership talent – the ability to build and maintain high-performing teams, and to inspire followers to set aside their selfish agendas in order to work for the common interest of the group.' So, as a result, 'The paradoxical implication is that the same psychological characteristics that enable male managers to rise to the top of the corporate or political ladder are actually responsible for their downfall. In other words, what it takes to *get* the job is not just different from, but also the reverse of, what it takes to *do the job well*.' Maybe, instead of sending women on assertiveness training courses, we should send men on humility and bullshit-avoidance courses, and the authority gap might be better addressed.

Still, over-confident applicants are more likely to be offered a job, even when their self-esteem bears no relation to their ability.[22] Boys and men can bluster and swagger and take up disproportionate physical and conversational space, and people will assume that they know what they're talking about. They can self-promote and people will believe them. Yet it's not just that women tend to feel uncomfortable doing this; they can't get away with it even if they attempt it. Try to create a mental image of a woman swaggering and take a quick moment to assess how uncomfortable it makes you feel.

So should women 'lean in' more? Should they be more confident, assertive and demanding? In other words, should we blame the authority gap on women? If only they were more outwardly confident, would it go away? Sadly, things are much more complicated than that. For when women behave as confidently as men, we make them suffer.

Women are often blamed for not being assertive enough. At the extreme, we even hear people claim that the gender pay gap is all women's fault: if only they asked for more money, they would earn as much as men. But researchers in Australia found that women *do* ask for a pay rise just as often as men. They are just not given it.[23] Women are often punished for being as assertive as men. They don't get the pay rise, the promotion or the job, even if they ask for it.[24] And this is because women, unlike men, are rewarded for being likeable.

We – and particularly men – want women to show what social psychologists call *communality*: kindness, warmth, unselfishness and nurturing qualities, which are stereotypes associated with the female gender. They tend not to like women showing *agency*: determination, decisiveness, assertiveness and leadership traits, which are stereotypes associated with men. Yet confidence and assertiveness are all about agency.

A survey of MBA graduates found that just as many women as men tried to negotiate a higher salary than they were first offered for a job, but men won higher offers than women.[25] And women who do negotiate are taking a big risk. Female job applicants who negotiate are twice as likely not to be hired as men who negotiate.[26] Women are penalized more than men for asking for more money, and potential hirers are more than five times more likely to say that they don't want to work with a woman who negotiates than with a man who does. This bias comes entirely from men: male hirers don't dislike men who negotiate, but they do dislike women who do. Female hirers treat both genders the same.

There is a double-bind for self-promotion too. Because women are often assumed to be less competent than they are, if they don't highlight their successes and achievements, they will lose out on jobs and promotions, and will be undervalued by their colleagues and bosses. If they do self-promote, though, they will be disliked – and also, therefore, miss out on jobs and promotions. For nobody likes a boastful woman, and women have to be liked to be hired. Men suffer none of these biases. They can be confident and people will still like them. They can promote themselves and people will both take them at their estimation and not see them as boastful. They can negotiate on their own behalf and this will be seen as perfectly normal and acceptable.

The sociologist Laurie Rudman has put this to the test.[27] She found that in settings where self-promotion was important for getting hired or promoted, women who behaved confidently and assertively were much less well received than men. They were generally thought to be more competent, but also less likeable and hireable. So if women don't 'lean in', as advised by Sheryl Sandberg, they won't stand a chance.[28] But if they do, they may be disliked and still not hired. And, because women are expected more than men to be likeable, this becomes a triple bind.

It is very hard to counteract these biases, but there are things we can do to mitigate them. If we are considering hiring or promoting people, we have to be incredibly careful to make our judgements based on concrete evidence, not on our intuitions about them. So, for instance, writing out a very clear list of specifications and marking each candidate rigorously against it can help to prevent us hiring a man because he seems more confident or offering him the job based on his potential but judging a female candidate only on her past achievements. We should take people not at their word – that allows men to inflate their credentials – but on what they have actually done. And we shouldn't ask questions in interviews such as 'What are you most proud of?', as this will enable men to boast but will simply make women feel uncomfortable, as they don't have the same licence to self-promote. Finally, we have to resist letting likeability sway our decisions.

Blaming women for not being confident or assertive enough and just telling them to 'lean in' is far too simplistic. As Deborah Cameron writes: 'Instructing women to behave more like men (interrupt more, smile less, stop apologizing, etc., etc.) takes no account of this evidence that women are judged by different standards. They are caught between a rock ("your speech lacks authority and no one listens to you!") and a hard place ("you're too abrasive and no one likes you!").'[29] Women have to try to combine confidence and warmth in exactly the right ratio for them to be taken seriously but not disliked in the process. (And it's no use saying that they should grow a tougher skin and stop caring about being disliked. For, unlike men, they have to be liked in order to be hired, promoted or listened to.)

In order to understand the double standards that are applied to women and to men, it may help to enumerate the adjectives that are used about successful, confident women, but not men: 'bossy', 'abrasive', 'strident', 'shrill', 'aggressive', 'scary', 'cold', 'stern', 'controlling', 'ball-breaking', 'bitchy', 'unlikeable', 'pushy', 'ambitious'.

'Ambitious', when applied to women, is particularly interesting. It was used of me a lot when I was younger: 'Mary Ann is an assistant editor at *The Times* and has been called ambitious so many times that her children probably think it is her middle name,' read a profile in the *Independent*.[30] Yet you don't get to the top of any organization without being ambitious. It's so taken for granted in men that it's rarely even noted, and certainly not in the derogatory way in which it's used against women.

Elaine Chao was described in the liberal *New York Times*, of all places, as '*unapologetically* ambitious' [my italics].[31] What exactly was she supposed to apologize for? She was a very senior politician who served two terms as Labor Secretary and was, at the time of writing, in her third Cabinet term at Transportation. 'I actually thought it was very strange,' she replied, when I asked her about that description. 'I am still ambivalent about using the word "ambitious", especially given the fact that I'm Asian, and Asian culture is very modest, very humble. So, "ambition" is a Western word that connotes self-centredness.'[32]

Baroness (Helena) Kennedy, an eminent human rights lawyer, told me about a conversation she had with one of the most senior judges in the UK after three men were chosen to fill all three vacancies on the Supreme Court. At that time, Baroness Hale was the only woman out of twelve on the Supreme Court, and it was surely time, Kennedy thought, for at least one more. She took this senior judge to task. 'He said, "Oh, there just weren't the quality of applications," and I said, "Oh, really?" And I mentioned a particular woman, to which he replied, "She's too pushy." You just know that would never be said about a man. I was utterly shocked.'[33] What's more, this male judge clearly saw nothing wrong in eleven out of twelve justices being male (and white) in a court that made decisions that would affect the whole British population. Surely a Supreme Court should be just a little bit more representative of the country it passes judgement upon?

If 'pushy' is seen as unacceptable in a woman, power-seeking is even worse. Bernardine Evaristo has often come up against this, she told me. 'I was talking to a female friend about the fact that I'm very much a literary activist, and I like to make a difference and see things change in our society. And I said, "I like to be powerful." And she was so shocked. She questioned me and was almost disgusted at the fact that I would want to be powerful.

'You know, we should all want to be powerful, obviously using power for the greater good, but there's nothing wrong with power in itself. It needs to be shared. And yet she was so disparaging of the fact that I wanted to be powerful. And then I was interviewed not so long ago by a guy. And I was talking about ambition and how you should go for what you want and so on, and he said, "Oh my God, you're power mad!" He was joking. But he wasn't really.

'What does "power mad" mean? And why are we as women accused of being power mad when we're just being strong and powerful? There's nothing mad about it: we want to have agency, don't we? We want to have a say in how the society is run. We want to be a voice and to be heard, and to make things happen. And in order for that to happen, we have to be powerful. So it's something to aim for. And it's positive and it's good and it's great. It's only if you abuse your power that that's a negative thing.'[34]

We have to find a way of admiring and cheering on powerful women so that they can do good in the world. While we continue to feel queasy about them wielding authority, the gap will never close.

Like Evaristo, Muriel Bowser, mayor of Washington DC, labours under the twin disadvantage of being black and female. She's well aware of the double standards women leaders face in politics, particularly if they are executives, like mayors and governors, who often have to make painful decisions, rather than legislators who only have to make speeches and pass laws. 'People don't like women promoting themselves,' she told me. 'It has a lot to do with how people see women in executive roles. People aren't used to seeing women in executive roles where you have to make quick decisions and there are winners and losers, because women are supposed to make sure that nobody loses. And that's not possible when you are the executive and the buck stops with you. Men are seen as decisive, whereas you are supposed to work longer, talk softer, never use curse words, all of those things.'[35]

The older women get, the easier they often find it to be assertive or confident. Elaine Chao told me how she was prepared to stand up to President Trump. 'He respects me. I'm not afraid of him. He smells fear. Maybe because I am the age I am, I've done this job before, I know what I'm doing. And so, I'm not afraid to approach him, I'm not afraid to call him, I'm not afraid when he calls me. And I give him my unvarnished opinion. I'm tactful and diplomatic. I select a good time. But I'm straight with him, and he respects that.'

Often women simply have to stand up for themselves, even in the face of male bullying. Mary Robinson was President of Ireland when Charles Haughey was Taoiseach, or Prime Minister. She told me a great story: 'He wasn't happy – as he saw it – that I was exceeding my constitutional powers as President: doing too much, basically, having too many meetings, seeing too many people, going round the country too much, going abroad too much, and so he got a legal opinion to put me back in my box. We met privately in my study and we argued, and I'm a constitutional lawyer, so I got the better of the argument. Every point he raised I won, even though he was a lawyer himself. And eventually he threw the legal opinion on the floor and said, "Oh, lawyers, you get what you pay for!" I was the one smiling as we came out the door, and I never got that kind of problem again.'[36]

Robinson was really sure of her facts, which enabled her to get the better of Haughey. This is a technique that my interviewees brought up again and again as a way of boosting their confidence. They daren't blag, in the way many men do; instead they put in hours and hours of preparation to be absolutely sure of their ground. And the preparation itself gives them confidence.

Janet Yellen is famous for the care she takes to get everything right. She even writes out, 'Hello. My name is Janet Yellen' at the top of her speeches. As she explained to me, 'A fundamental thing about me is that I do a lot of preparation. I don't wing it . . . I don't feel super-confident, no matter what position I've been in, including as Chair [of the Federal Reserve]. I've never personally developed a feeling of, "I'm on top of this, I can do anything, I can just relax because it will be assumed that I'm on top of all of this." '[37]

Her vice-chair, she told me, used to try to reassure her that all the hours of preparation she put in before press conferences weren't necessary. 'He said, "You could just walk into this thing. You don't have to do any work, you could do absolutely fine. You're spending much too much time on preparing for things I know you could just go in and do." And, that may have been true, but I felt better about it, and I needed to feel more confident.'

Dame Elizabeth Corley admitted to me that, 'I'm an insecure over-achiever. I'm constantly putting myself in situations where I think they're going to find me out, I'm going to completely fail, I'm going to fall flat on my face. People talk about impostor syndrome and assume it disappears as you become more senior – that you have security in your position. I think that's true as long as you don't push yourself on, but if you do, you just have to prepare really, really hard.'[38]

Because most women instinctively sense that there's a danger their work will be undervalued relative to men's, they have learned that they can't take risks. They have to be completely on top of their game. Elaine Chao outlined the strategy to me: 'Women tend to both develop a real body of expertise and just prepare, prepare, prepare. They just do their homework so much more conscientiously on the whole than their male colleagues. Because you can't let a crack appear, can you?[39]

'This is also why I think women are unfairly criticized for being automatons, for being too robotic. I'm much better than I used to be, in terms of being more spontaneous, being more able and willing to wing it. But I still like to be prepared, it just gives me an added sense of confidence, and comfort, so that I can perform better.'

Christine Lagarde and Angela Merkel have even compared notes about their tendency to over-prepare. 'We have discovered that we both have the same habit,' says Lagarde. 'When we work on a particular matter, we will work the file inside, outside, sideways, backwards, historically, genetically and geographically. We want to be completely on top of everything and we want to understand it all and we don't want to be fooled by somebody else. We assume somehow that we don't have the level of expertise to grasp the whole thing.

'Of course, it is part of the confidence issue to be overly prepared and to be rehearsed, and to make sure that you are going to get it all and not make a mistake . . . It's very time-consuming!'[40]

So these women have all found that they can boost their confidence by making sure that they are absolutely on top of their brief. Another useful piece of advice, from the first ever female Bishop of London, Dame Sarah Mullally, is a physical one. 'If I'm going into somewhere that I think is going to be tough, I pull my shoulders back, I will sit back in my chair, I will occupy the space. I know where I'll sit at a table. I do that very consciously because that's important, so I'm seen. And if it's a tough meeting, I'll wear red lipstick.'[41]

Women have to be very aware of the space they take up. Often they will lean forward at a meeting, seeing it as a sign of engagement. Men, though, see this as a sign of weakness, and are more likely to respect people who sit back in their chair, displaying confidence and strength. And because women are physically smaller than men, they have to make sure they are not exacerbating the difference by hunching their shoulders or shrinking into themselves.

Another tip comes from Anne Mulcahy, the rescuer of Xerox. She forces herself to take on difficult challenges. 'I fight the confidence battle all the time,' she admitted to me.[42] 'Not that I overly lack confidence, but I still get nervous: am I good enough? Someone once said to me that sometimes women are motivated more by the fear of failure than they are by the desire for success, which is interesting because it's probably characteristic of who I am.

'I fight it all the time. I make myself speak up. I absolutely ensure that I don't get reticent in situations where I'm uncomfortable, that

I do step up to the plate, that I take tough assignments. So I'm always in a battle with myself about fighting the demons of lack of confidence, and I think over time it's gotten easier.'

It certainly does get easier with age. But that is no consolation for younger women. My advice to my daughters is always to act confident even if they don't feel it. Then, at least people are likely to treat them with more respect, which in itself will boost their confidence. But that doesn't help with the problem of being disliked. A newspaper profile of me when I was a young(ish) journalist remarked that, 'Her confidence is legendary.'[43] After all we have read in this chapter, I fear that might not have been a compliment.

One explanation for the authority gap, then, may be that men tend to project more confidence in their views than women. This is because they have been socialized from childhood to blag, to boast, to speak up and self-promote, while girls have been penalized for exactly the same behaviour. Men have also been brought up to believe, on average, that they are cleverer than women, and that men can be brilliant, while women are just diligent.

Women, meanwhile, have internalized this bias. And even if they do feel confident, they have learned that it doesn't always pay to show it. Yet they will be penalized too if they are modest or self-deprecating, as they are more likely to be underestimated than men.

We have to learn to cut through this bullshit. Parents and teachers can try to correct the bias in the next generation, by making home and school as much of a confidence factory for girls as they are for boys. Girls should be rewarded for talent as well as diligence and encouraged to speak up in class. Boys need to be taught not to hog the floor and to have a realistic assessment of their abilities.

In this generation, though, boys and men are much more likely to overestimate their ability and girls and women to underestimate theirs. So instead of taking people at their word, we need to assess much more rigorously and objectively how good the women and men in front of us are. We shouldn't punish women for being 'under-confident' or 'over-confident'. Instead, we should understand that we make it much harder for women than for men to get this right.

Indeed, we need to learn to value the whole spectrum of behaviour: to listen just as carefully to the quiet contributor to the meeting as to the blusterer. Just because a man speaks out readily doesn't mean that he knows what he's talking about. And just because a man asks for a pay rise doesn't mean that he deserves it more than the woman who hasn't asked for it.

We should also check the words that come to mind when we encounter a woman with authority. Is she really strident, abrasive or a bitch? Or does that tell us more about ourselves than about her? Would we use the same adjectives of a man displaying the same behaviour? If not, we should reassess.

Personally, I find confident, sassy, ambitious women completely fabulous, but I fear that I'm in a minority, given that most men and some women disagree. In an ideal world, we would all see them that way. Until then, if we want to narrow the authority gap, we need to look very carefully at our instinctive reactions to people and try to correct them with a more rational overlay. Most of all, we have to stop mistaking confidence for competence. They are two utterly different things.

6

Conversational manspreading
How men hog the floor

'I'LL SAY IT AGAIN... IT'S LIKE THIS... I THINK
YOU'LL FIND... WOMEN TALK TOO MUCH '

'There is no form of privilege men deploy more frequently, more casually and more unselfconsciously than their assumed Divine Right to Talk – to monologue, to mansplain, to interrupt, to say whatever's on their minds without considering the consequences.'

– Deborah Cameron, professor of
language and communication

MAYBE IT'S BECAUSE I'm a woman or maybe it's because I'm a journalist, but if I sit next to a man at a dinner party I immediately start drawing him out. It's polite and it's a good way of instigating a conversation. What really irks me, though, is when (and it happens all too often) he doesn't ask me a single question in return.

Once, when I had exhausted every conversational avenue with the man on my right, and had discovered everything there was to know about his life, his career and his family over at least forty-five minutes, I finally lost patience. 'Well, it's been lovely hearing all about you,' I said, smiling sweetly, 'but according to conversational etiquette, now's the time to ask what do I do.'

'Ahhh!' he exclaimed. 'What do *I* do? Well . . .' and started expatiating about himself again!

Louise Richardson told me a similar story. 'A female vice-chancellor of another university told me she'd been at a dinner between two men, one was a head of college. She asked one or two questions and he spoke all the way through the first course, all through the main course and at the end of the main course, he said, "Well, that's enough about me," and he turned to the person on the other side!'

The best description of this is *conversational manspreading*, taking up too much conversational space at the expense of the people around you – usually a woman on each side. And it's not just rude to talk too much about yourself; it's also rude to show no interest in your interlocutor. What you're suggesting is that you are much more interesting than she is; though how you would know that is a mystery, if you haven't even bothered to engage with her in the first place.

This is both a manifestation of the authority gap and, in a sense, another explanation for it. If men talk more, they will be listened to more. Maybe if women spoke up more, particularly in public, they would be accorded more authority for what they say. But equally, if men bang on and don't leave any space for women, then it's hard for women's voices to be heard. And when women do speak up, they are often not let in – or they are disliked for being too talkative. Yet another double bind.

The author, journalist and broadcaster Bel Mooney put it beautifully in a message to me: 'I cannot tell you the number of times I sat at dinner tables with my ex-husband and politician friends. The men did all the talking. Oh, they talked, they talked. Their confident opinions rose to the ceiling like balloons and their faces shone in candlelight, full of the glory of themselves. If any of the wives squeaked up? A nod, a smile maybe . . . and then the talk would flow on, like the wake of an ocean liner, the female seabirds bobbing helplessly in the churning foam. Why? Because the successful male voices were the voices that carried, the opinions that mattered – while the intelligent wives were there to look elegant and help carry out the plates. From time to time the talk would move from politics to books, and I (who knew more about literature then any of them) would try again, but my tentative tones would make no dent in their great, masculine wall of sound.'[1]

The conventional wisdom is that women are chatterboxes, that they talk far more than men. In fact, if you wire up men and women for a day and count the number of words they've used, the total is almost exactly the same – about 16,000.[2] In that study, the three top talkers, using up to 47,000 words a day, were all men. (But so was the most taciturn, at 700 words.)

What's certainly true is that women talk less on average than men in public settings. And it's because they are doing the opposite of conversational manspreading, says Deborah Tannen, professor of linguistics at Georgetown University: 'One reason women tend to speak less at meetings, in my view, is that they don't want to come across as talking too much. It's a verbal analogue to taking up physical space. When choosing a seat at a theatre or on a plane, most of us will take a seat next to a woman, if we can, because we know from experience that women are more likely to draw their legs and arms in, less likely to claim the arm rest or splay out their legs, so their elbows and knees invade a neighbour's space. For similar reasons, when they talk in a formal setting, many women try to take up less verbal space by being more succinct, speaking in a lower voice and speaking in a more tentative way.'[3] We're back to Kate Manne's explanation of male entitlement. (Some) men feel entitled to hog the conversational time, just as (some) men feel entitled to hog the leg space on the Tube.

As Mary Beard put it to me, 'You don't find a man saying, "Oh God, did I talk too much?" Whereas, "I'm sorry, did I talk too much?" is a classic female phrase. Why it's difficult to address is that every individual instance seems trivial, but the aggregate is very important.'[4] Men feel entitled to take up conversational space, while women are embarrassed when they do it. It's as if they are only around the table on sufferance and need to apologize for their presence.

In a classic study of this phenomenon, Barbara and Gene Eakins recorded seven university faculty meetings.[5] They found that, in all but one case, the men at the meeting spoke more often and, in all cases, they spoke for longer. The longest comment by a woman at all seven meetings was shorter than the shortest comment by a man.

And it was true, too, in the Trump Cabinet, according to Elaine Chao: 'I think the guys talk more. I try not to say anything unless it's really important. And I'm sure some other guys just want to vent, because they want to vent. Bless their hearts, they really do think they're contributing.'⁶

Sue Montgomery, a mayor in Canada, put this to a clever visual test. Because she likes to knit during city council meetings to keep her mind focused, she decided to change colour between red and green wool every time a man or a woman spoke. 'I would say it's probably 75 per cent, 80 per cent red and the rest, there's little bits of green here and there,' she said.⁷ This isn't because men make up 75 to 80 per cent of the council; there are thirty-one female council-lors and thirty-four male ones. So why the gap? 'The women are much more efficient, stand up, make their point, sit down,' she said. 'Men like to hear themselves talk. Not all men, but there's a handful of men who take up a lot of space.' You can see a picture of her shawl online.

Men also like to make themselves heard after other people speak. A study of 250 academic seminars in 10 countries found that men were two and a half times more likely to ask a question than women.⁸ When the researchers followed up by asking men and women how frequently they asked questions and why, they found that women were much more likely to say that they didn't have the nerve or that they were intimidated by the speaker. They also found that, if a woman asked the first question after a talk, more women would fol-low her. If I am chairing a panel or a talk, I make a point of calling a woman for the first question in an attempt to even up the numbers.

Sometimes there are cultural factors that hold women back from taking the floor. Dina Kawar is a fantastic role model for younger women in the Middle East. She was the first Arab woman to preside over the United Nations Security Council as Jordan's am-bassador to the UN. She is impressively articulate in English and looks supremely poised. Yet she told me, 'The worst struggle for me was to struggle with myself as a female because there are taboos that you grow up with that never go away. The self-inhibition is the hard part. When there's a discussion, I had to learn to speak out and not be always the last because [I had been taught that] the men had to speak first.'[9]

For men in the Middle East, she says, 'When a woman talks, it's always at the very beginning, "Oh, she's cute." It takes time for them to start to say, "OK, she's worth listening to." Luckily, I'm over it now,' she laughs.

There is evidence that if women are exposed to female role models, they will speak for longer.[10] Young women and men were asked to give a speech while being subtly exposed to a picture of either Hillary Clinton, Angela Merkel or Bill Clinton or none at all. Women spoke less than men when there was a picture of Bill Clinton or no picture. But when they saw Hillary Clinton or Angela Merkel, the gender difference vanished. And they spoke more eloquently too, as judged by the listeners. It makes you wonder about the subliminal effect on female students at universities stuffed almost entirely with portraits of men.

For what we see around us really does make a difference. Even Baroness (Minouche) Shafik found it dispiriting when she was Deputy Governor of the Bank of England and 'there were all these portraits on the walls, these huge, big oil paintings of just one old white guy after another. There was just one small portrait of a woman who had been my predecessor as a deputy governor, but that was it, there were hundreds of portraits, corridors of these old white men.'[11]

My favourite study about speaking time – because it makes me laugh – involved showing men and women three paintings by Albrecht Dürer and inviting them to talk about the pictures into a tape recorder for as long as they wanted.[12] The women spoke on average for 3.17 minutes and the men for 13 minutes: in other words, four times as long. But these statistics aren't entirely accurate, because three of the men were still talking when the tapes, which had a recording time of thirty minutes, ran out!

I wrote in Chapter 5 about how boys are encouraged to hog the floor in classrooms. Teachers ask more questions of them, call them up in front of the class more often and reward them for talking, while praising girls for being quiet. David and Myra Sadker and Karen Zittleman, who observed this behaviour, tell a story that illustrates the effect of this perfectly:

> David [Sadker] recalls a unique student gathering in an auditorium in a Midwestern high school. More than a hundred school newspaper reporters were gathered for a gender equity 'press conference' to ask David and others about their research. At first the students seemed reluctant, but then the comments came quickly, an avalanche of questions.
>
> 'Hold it a minute,' David said, halting the rapid-fire pace of the press conference. 'Do you notice anything ironic going on here?' The room of reporters looked around blankly. Then a girl from the back, where most of the females were clustered, said, 'The boys are asking all the questions.' There was an audible gasp as the students realized they had become living proof of the story they were supposed to report.

As the authors write: 'Sitting in the same classroom, reading the same textbook, listening to the same teacher, boys and girls receive very different educations. From grade school through graduate school, female students are more likely to be invisible members of classrooms. Teachers interact with males more frequently, ask them better questions, give them more precise and helpful feedback, and discipline them harshly and more publicly.

'Over the course of years, the uneven distribution of teacher time, energy, attention and talent shapes both genders. Girls learn to wait patiently, to accept that they are behind boys in the line for teacher attention. Boys learn that they are the prime actors shaping classroom life . . . In today's sexist school culture, boys assume that they are number one, and begin to understand the inherited power of entitlement.'

And, as we learned in Chapter 3, when boys grow up to be men, many of them act out that entitlement by interrupting women, talking over them, ignoring them, resisting their influence, challenging their expertise, taking up more conversational space, discounting what they have to say, and mansplaining to them even when the women themselves are experts. No wonder many women have concluded that there is little point in speaking up.

As the philosopher Kate Manne says in *Entitled: How Male Privilege Hurts Women*, 'When it comes to knowledge, especially of a prestigious sort, the idea that men have a prior claim to it is as venerable as the patriarchy itself. Sometimes it's connected to the idea that women are incapable of being authority figures. In *Politics*, for example, Aristotle wrote: "The slave is wholly lacking the deliberative element; the female has it, but it lacks authority."

'Part of what's going on is the presumption that a woman will be less knowledgeable, less competent and somehow in need of a man to explain things to her. That doesn't explain the fact that mansplaining often also involves men's resistance to evidence that the woman is more knowledgeable on the subject than he is, and sometimes, the anger when that turns out to be the case.'[13]

Women have good reason to wait until they're certain of their knowledge before speaking up, because, if they're not, they may be penalized. Catherine Tinsley and Robin Ely, professors of management and business administration, saw a classic example of this dynamic at a biotech company.[14] The female research scientists talked far less in meetings than their male colleagues did, yet afterwards, in one-to-one conversations, often offered insightful thoughts. What their bosses had failed to notice was that when the women did speak in meetings their ideas tended to be shot down if they contained even the slightest flaw. In contrast, when the men's ideas were flawed, the best elements were rescued from them. Women therefore felt they needed to be absolutely sure of their ideas before they would dare to share them.

But if men in a meeting *have* to listen to a woman's opinion, it's amazing how much difference it makes. For their book *The Silent Sex*, Christopher F. Karpowitz and Tali Mendelberg put men and women in groups of five to come to a decision about how a locality should raise and spend money.[15] The groups contained every possible permutation of men and women. In almost all the groups, the men talked disproportionately more and the women disproportionately less. The only two circumstances in which women talked for a proportionate amount of time were either if they made up at least 80 per cent of the group – i.e. there were four or five women out of five – or if the decision-making process was changed from a simple majority to unanimity, like a jury. When every member had to agree in order for the group to come to a conclusion, women calculated that their views *had* to be listened to and taken into account. And so they spoke up.

But even increasing the numbers of women in a deliberative group doesn't necessarily allow them to increase their influence. Sometimes it leads to a backlash. For example, when there was a large influx of women MPs into the House of Commons after the Labour landslide in 1997, they were the target of horrible sexist behaviour from some of their male colleagues, including Conservative MPs putting their hands out in front of them as if they were weighing melons when a woman stood up to speak.[16]

It's not just that women don't think they'll be listened to if they speak more. It's that they know people don't *like* them speaking as much as men. For millennia, men have tried to silence women, and they are still at it, as we'll see in Chapter 14.

Women also suspect that, even if they do take up only proportionate conversational space, listeners will feel that they have talked more. And they're right. When people in a study were asked to listen to identical scripted conversations between a pair of women and a pair of men and each spoke for exactly the same length, the listeners calculated their speaking time accurately. But when the very same dialogue was acted out by a man and a woman, both male and female listeners thought the woman had talked more than the man, even though she hadn't.[17]

Women know this instinctively, as Helle Thorning-Schmidt has noticed: 'I think you can go to any meeting and time people (and I do that sometimes), and men will speak much longer than women. But if you ask people afterwards who spoke the most, they will either estimate that they spoke more or less the same, or they will say women spoke longer. It's a successful strategy for men to speak for a long time, but it's not a successful strategy for women and then they stop doing it.'[18]

Surely powerful, authoritative women don't have this problem? Don't they feel that their power gives them licence to dominate a room? This is true for men, but not for women. Victoria Brescoll of the Yale School of Management studied two congressional sessions in the US Senate and measured the amount of time each senator spoke.[19] She found that the more powerful a male senator was (measured by how long he had been in the Senate, whether he chaired a committee and so on), the more voluble he was, the more he spoke. But this wasn't the case for female senators. They didn't talk more when they were more powerful.

Brescoll wanted to know why this was. Were powerful women not taking advantage of their power because they preferred to lead in a more democratic and less hierarchical way than men? Or was it because they feared a backlash: that the more they acted in a traditionally male, dominant way (as befits a leader), the more other people would dislike them? This is what social psychologists call the *status incongruity hypothesis*: because women are assumed to be subordinate to men, if they behave in a powerful way, their behaviour is incongruent with their gender, so it makes both men and women feel uncomfortable. It is why we punish women for being agentic (assertive, dominant, confident, taking charge), because that sort of behaviour is traditionally male.

To try to find out, Brescoll conducted an experiment. She asked 206 men and women to imagine that they were in a marketing meeting of four people trying to come up with a strategy. Half were told that they were the most powerful person in the room; half that they were the least powerful. She then asked them questions about how much they would talk in the meeting, to what extent they would try to establish a rapport with the others, and whether they worried about the others making judgements about their talkativeness.

The pattern was similar to the one in the Senate. More powerful men said they would speak for longer than less powerful men, but more powerful women said they would speak for the same time as less powerful women. There was no relation between volubility and wanting to establish a rapport, suggesting that this isn't a question of women having a more democratic leadership style than men.

There was, though, a strong relationship between talking time and fear of backlash. This was measured by asking questions such as: 'Would you be concerned that you might be disliked?'; 'Would you be concerned that people would judge you for how much you talked?'; and 'Would you worry that people thought you dominated the meeting?'. The more the powerful women agreed with these questions, the less they said they would talk. But men didn't fear a backlash if they talked for longer.

Is women's fear justified? Brescoll went on to test whether people think less of powerful women if they talk a lot. (I initially wrote 'talk too much', before realizing that this in itself illustrates the problem. Is 'too much' different for women and men?) She asked men and women to rate a fictitious CEO who talked 'much more than others in power' (as CEOs often do).[20] When the CEO was given the name Jennifer, both men and women rated her as much less competent and less suited to leadership than when they were told that the talkative CEO was called John. When 'Jennifer' was described as talking less than others, her perceived competence and leadership suitability shot up.

Louise Richardson believes that women learn to have more emotional intelligence in meetings precisely because of these problems. 'It ends up making you a more nuanced thinker, a more self-aware person in your interactions. I pay much more attention to the impact I'm having in conversations with people around me, which I think men don't do. They come in in broadcast mode, they've got something important to say and they say it. Women can be so much more successful when they get to the top, because they've just acquired these skills, they're in their DNA as a result of navigating up.'[21] Women can tell quickly if their interlocutors are getting impatient or are glazing over and they learn to adjust their style accordingly.

It's the lack of this self-awareness that is so striking in the minority of men who just hog the floor regardless. Richardson gave me a perfect example. 'I occasionally pull out my phone and surreptitiously time how much of a conversation the man I'm with is taking up. I have someone who's subordinate to me and I used to time him. He would come in and just speak *at* me. He no longer works with me; I didn't renew his contract. I asked, "Do you not think maybe you might be interested in what I think about what you're telling me, or how I react to what you're telling me?" It's not even smart. I was his boss, after all!'

Yet again, women face an almost insoluble dilemma. Talk as much as men and they are deemed to be talking 'too much' and are then seen as less competent and less likeable. Talk too little and they will have no influence or authority. Most senior women have learned to say exactly what they need to say, no more and no less, in order to be taken seriously. They have to be confident enough to cut through the barrage of male voices but warm enough not to alienate them. It's as if women need the discipline and agility of an Olympic gymnast to stay on the balance beam of acceptable conversation, while men can just saunter across the floor.

It may not just be a question of how much women talk, but also the sound of their voice that detracts from their authority. The *Channel 4 News* presenter Jon Snow believes that 'It's bass registers that give authority to a voice.'[22] Other men whom I have talked to about the authority gap have echoed him. So maybe we accord less authority to what women say partly because of the higher pitch of their voices?

When a woman, Vicki Sparks, commentated on a 2018 men's World Cup football match on the BBC, you can imagine the complaints about her voice. It was 'too high-pitched' and a 'tough listen' for footballer Jason Cundy. Others called it 'squeaky', 'screeching', 'shallow', 'shrill', 'strident' and 'annoying'.[23]

'Contrast the "deep-voiced" man with all the connotations of profundity that the simple word "deep" brings. It is still the case that when listeners hear a female voice, they do not hear a voice that connotes authority; or rather they have not learned how to hear authority in it,' writes Mary Beard in *Women and Power*.[24] Do we associate a deep pitch with authority because it is genuinely more authoritative or because we associate a deep voice with 'male' and we associate 'male' with 'authority'? These biases are so intricately woven in our psyches that it is impossible to disentangle them. Like so many aspects of the authority gap, our unconscious brains trick us by using heuristics rather than reason. Think 'male', think 'leader'. Think 'deep', think 'authoritative'.

Women know what a handicap their voices can be to being taken seriously. So it's fascinating that the average pitch of a woman's voice has fallen significantly over the past few decades in more egalitarian countries.[25] You only have to watch a black-and-white film made in the 1950s to notice how much higher women's voices were then.

And the effect is stronger the more egalitarian a country is. American women's voices are lower than Japanese ones, Swedes' are lower than Americans', and Dutch women speak lower than Swedish women.[26] In Dutch society, which is fairly androgynous, there isn't much difference in pitch between men and women. In Japan, by contrast, women use the higher ranges of their voices much more than in Western countries.[27] By doing this, they are demonstrating highly traditional feminine traits such as submissiveness, powerlessness, deference and subservience. When Japanese women are being polite, they can reach an abnormally high peak of 450Hz, while English women in one study never exceeded 320Hz. Japanese men, meanwhile, speak at a lower pitch than English men, even though they are physically smaller.

When Theresa May, Britain's second female prime minister, was asked what advice she would give to an aspiring young woman entering politics, she replied, 'Behave like the men.'[28] Britain's first female prime minister, Margaret Thatcher, realized she had to sound like them too. Her advisers thought that she sounded 'shrill' in the House of Commons – always a danger for a woman – and lacked authority. She ended up lowering her voice by a full 60Hz, or half the difference between an average female and male voice. It never sounded quite right, though. It had an artificial and condescending tone to it: the journalist Keith Waterhouse once wrote, 'I cannot bring myself to vote for a woman who has been voice-trained to speak to me as though my dog has just died.'[29]

Dame Margaret Hodge is a veteran Labour MP. She has held several ministerial jobs and has chaired the House of Commons Public Accounts Committee. What she has learned, she told me, is that 'in the House of Commons, your voice matters a lot. If you're in a debating arena and you're a woman, it becomes high-pitched. Anne Campbell, the Cambridge MP, had quite a high-pitched voice. And if she ever tried to raise it, she would then be laughed at. It became a way of undermining the credibility of what she was saying. The atmosphere is rowdy. A woman comes in with a voice which tends to be higher-pitched, and that gets attacked and nobody listens to what she says. I am always really careful to keep my voice low. It's something I've learned over the years.'[30]

The German chancellor Angela Merkel agrees. As she told *Die Zeit*, 'A woman's voice is not as dark and strong as a man's voice. For a woman, radiating authority is something you have to learn.'[31]

Baroness (Helena) Kennedy, a human rights lawyer, told me that she often talks to women who have applied to become judges but have been told that they don't have sufficient authority. 'So what does authority mean? What is it we're looking for when we talk about authority? Is it that we talk in a rather sonorous voice and we can be patronizing to people and we are able to put our stamp on the court? It's because authority is linked in the minds of those who make appointments with a certain, male way of being. Yet the stamp of authority comes from the very status you're given as a judge. So measuring authority before you become a judge is very difficult. Because sitting on a dais above everyone else wearing a robe gives you a stamp of authority.'[32]

Mamokgethi Phakeng, vice-chancellor of the University of Cape Town, has come across exactly the same phenomenon. 'The one critique that I've heard is that people say, "She lacks executive gravitas." They say this in selection committees whilst we are recruiting for leadership positions in the university. And I say, "What do you mean?" And people say, "She doesn't sound believable." Nobody says that about a man.'[33] In other words, her voice doesn't convey authority.

In one experiment, when played voices of different pitches, people rated the lower-pitched voices (both male and female) as more competent and more trustworthy.[34] It clearly pays for women to lower their voices if they want to be respected and trusted – and elected. Other things being equal, we are more likely to vote for candidates with a lower voice, even when the office is a traditionally feminine one, such as running a PTA.[35]

Joey Cheng of the University of Illinois tested this bias by asking small groups of people to discuss the items that an astronaut would need to survive a disaster on the moon.[36] At the end, she asked each one privately to rank the others according to their dominance in the group. What she found was that most people shifted the pitch of their voice in the first few minutes of the group discussion, and that those who lowered their pitch ended up higher in the pecking order, and vice versa. The lower-pitched ones were rated by others as being more domineering and more willing to impose their will over others and, as a function of that, they were able to gather more influence and make decisions on the group's behalf.

Women often know this instinctively. The Chinese-American designer and architect Maya Lin is physically small and has always looked much younger than she is. When she was just twenty-one, and still an architecture student at Yale, she won the (anonymized) competition to design the national memorial to the Vietnam War in Washington DC. The older white men who were in charge of putting the design into effect were not only astonished to find that she was young, female and Asian: 'the trifecta', as she puts it.[37] They also tried to overrule her wishes. She fought back tenaciously and, although she was tiny, had hair down to her knees, and 'looked about twelve', she had a low voice. 'I always thought, thank God I have a low voice, because I had to command people. For years, I was usually the only woman in the room and so having a low voice was my defence.'

It *is* a problem that women with high voices can sound childish in a way that men, whose voices break in puberty, can't. So women may be taken more seriously if they lower their voices a little, but getting the timbre right can be tricky, as Roula Khalaf, the first woman to edit the *Financial Times*, explained to me: 'Somebody told me, long before I became editor, that I had to project my voice more, that I was too soft-spoken, and that it wouldn't be good for my career. I think what he essentially meant was that I wasn't a man. That's how I took it. He wasn't used to not having a man in that particular position.'[38]

Being soft-spoken is equated with being hesitant and weak. For the same reason, women – and young women in particular – are often criticized for using 'uptalk', for going up in pitch at the end of a statement as if it were a question? It can be incredibly annoying to listen to, and it can be undermining of the speaker's authority, but it has its purpose. By turning a statement into a question, it is inviting the listener to listen actively, to nod or confirm. It can also help to prevent other people interrupting by suggesting that there's more to come. But it's not just a female thing. Men do it just as often as women; they just don't get criticized for it so much.[39]

There is also a danger in making your voice less soft-spoken. For what happens if women raise their voices? They become shrill, of course. There is nothing good about shrill. Men, by contrast, raise their voices and merely sound commanding. Hillary Clinton came up against this time and time again when she was running for President.

Nicholas Subtirelu conducted a media analysis during the 2016 presidential campaign and found that, although women were mentioned rather less than men in the US media, they were more than three times more likely than men to be accused of shrieking, and more than twice as likely to be 'screeching' or 'shrill'.[40] These were words used constantly of Clinton.

As Clinton has written, 'After hearing repeatedly that some

people didn't like my voice, I enlisted the help of a linguistic expert. He said I needed to focus on my deep breathing and try to keep something happy and peaceful in mind when I went onstage. That way, when the crowd got energized and started shouting – as crowds at rallies tend to do – I could resist doing the normal thing, which is to shout back. Men get to shout back to their heart's content, but not women. Okay, I told this expert, I'm game to try. But out of curiosity, can you give me an example of a woman in public life who has pulled this off successfully – who has met the energy of a crowd while keeping her voice soft and low? He could not.'[41]

We're back to the status incongruity problem. When we accuse women of being shrill, screechy, abrasive, strident, bossy, hectoring, grating or harsh, what we often really mean (though we may not know it) is that we feel uncomfortable with them exerting authority. It grates in two ways. We may dislike how they are going against type: it doesn't seem feminine to have strong opinions, to be as confident as men, to talk assertively, and to wield power. Or we may find them *over*bearing, which almost literally means they are displaying *too much* authority. If they were male, would we find them overbearing? Are our stereotypes getting in the way of us accurately assessing the person in front of us? Are we seeing her through a distorted template? Is our dislike more about us than about her? These are the questions we should always ask ourselves before passing judgement on an authoritative woman.

And we also need to calibrate more carefully our responses to people's volubility. Just because someone talks a lot doesn't mean that they have anything interesting or important to say. The reticent person at a meeting may be the wisest. Women have been taught from childhood not to speak out, as we have seen in observations of classrooms, and they have learned that they may be punished for talking 'too much'. We need to take that into account before we judge them.

We can also make efforts to bring quieter women into the conversation at a meeting and to ensure that men are not dominating the proceedings. We can call on a woman to ask the first question in a Q&A, as it will embolden other women to join in. And men can make themselves more aware of the conversational space they are taking up. Are they sharing it equally with a female interlocutor? Are they asking her as many questions as she is asking them?

Finally, we need to resist being seduced by a sonorous bass voice or put off by a higher female one. It's hard, but we have to learn to focus on the content of what someone is saying, rather than the pitch of their voice. As media organizations like to say: 'Content is king' – or maybe it should be 'queen'?

7

Changing our minds
How hard it is for women
to exert influence

'THANK YOU, MISS KHAN — THAT'S GIVEN US
A LOT TO SMILE INDULGENTLY ABOUT BEFORE
MOVING ON '

*'A weapon men use against women
is the refusal to take them seriously.'*
— David Mitchell, novelist

WE HAVE SEEN how men tend to come across as more confident and take up more conversational space. And we have all noticed how a woman can make a point and be ignored, only for a man to be acclaimed when he echoes her later. Combined with our tendency to underestimate women's expertise, these all contribute to a big difference in how much influence women have, compared with men. We are more likely to be swayed by a man than by a woman. And when I say 'we', I'm talking about women as well as men, though the phenomenon is particularly strong in men.

Kathleen Propp, from Northern Illinois University, put undergraduates into mixed-sex groups, supposedly to make a recommendation to a judge about a custody battle.[1] Some of the relevant information – 'Mother and father's marriage was problematic from the outset'; 'Father's mother-in-law believes that incest has occurred'; etc. – was given to the whole group, some to two members, and some to just one. They then had to come to a decision about custody, based on the information offered by different members of the group.

She found that information offered to the group by a man was twice as likely to be used for the group's final decision as information produced by a woman. If the information was known only to one person, it was six times more likely to be used by the group if introduced by a man. In other words, groups were far more likely to pay attention to and use information offered by a man but to ignore it if it came from a woman.

Men were much more influential than women. And if women's views are dismissed in favour of men's, it can have devastating consequences in the real world. It's why rape is still under-reported and under-prosecuted. It's why domestic violence was for so long ignored by the police. It's why childcare didn't become a political issue until there was a critical mass of female MPs. And it's why men have been allowed to get away with sexual harassment for so long.

Even being an expert doesn't always help women have more influence; bizarrely, it may sometimes make them less influential. One study took 143 business studies students, roughly half and half male and female, and gave them information about how to survive a bushfire.[2] They were then asked to write down, in order of importance, the items they thought would best help them survive. These individual rankings were compared against the official rankings drawn up by real experts, and the students who scored highest were deemed experts. They weren't told their scores, so they didn't know that they were experts, and the groups they were in didn't know it either. Women and men scored equally well on average, and women were just as confident as men of their ability.

They were then put into small groups, with one expert in each group, and asked to come up with a group ranking. The researchers measured how influential each person was by how much they had converted the group to their point of view. As you might expect by now, women were less influential within their groups than men. But, more surprisingly, female experts were perceived as less expert and were less influential than female non-experts. The male experts, meanwhile, were significantly more influential than the male non-experts.

How can this be? It doesn't seem to make sense. Well, it turned out that both women and men in the groups expected women to be less good at this task (even though they weren't) – a classic case of unconscious bias leading to underestimation. So the groups started the exercise with a disinclination to believe that the women would know what they were talking about. Then the experts often had to disagree with others during the group discussions in order to persuade them of their case. If they were women, this didn't go down well with the group, as people don't like women challenging and disagreeing with others (though they are happy for men to do so). The female non-experts, meanwhile, tended to agree with the group, which meant that they were rated as more competent and likeable – and therefore influential – than the female experts. None of these effects was seen in the men.

In another study, Ethan Burris from the University of Texas asked teams to make strategic decisions for a bookstore.[3] He randomly told one member that the bookstore's inventory system was flawed and gave that person data about a better approach. When that person was a woman, and she challenged the old system and suggested a new one, team leaders viewed her as less loyal and were less likely to act on her suggestions. Even when everyone in the team knew that one member knew something extra that would benefit the group, suggestions from women with inside knowledge were discounted.

This is what often happens in real life too. Women who challenge have a really hard time and face pushback. Frances Morris, Director of Tate Modern, has encountered this throughout her career, as she explained to me: 'For many years, I think I was often what you would call a disruptor. I would be the person who would say something that would change the pace of a meeting or be slightly contrary and occasionally this would be a brilliant idea. But it was almost always taken as a bad thing. I noticed that when male colleagues would do similar, it was almost always taken as a rather brilliant thing. And I've once or twice been taken to task by senior colleagues about that kind of behaviour. It's expected that if you are a female colleague you should be more compliant, more collegiate, more collaborative and hold back. When I look back on my career, I think it was that disruptive element that actually held back my promotion. For a long time, it was seen as inappropriate behaviour, but it wouldn't have been inappropriate from a chap. And that makes me sad.'[4]

Being challenging is part of the job description of being a TV political editor: you're there to hold the powerful to account. Yet Beth Rigby, political editor of Sky, is often criticized for doing so, she says. 'When I ask a challenging question, I'll often find men on Twitter saying things like: "She never lets up, does she?" There is an element of sexism and even misogyny there. Men ask challenging questions. Women are just nagging.'[5]

And as Mary Beard puts it in Women and Power, 'Unpopular, controversial or just plain different views when voiced by a woman are taken as indications of her stupidity. It is not that you disagree, it is that she is stupid: "Sorry, love, you just don't understand."'

Of course, in the bushfire study, the group didn't know that the women were experts. Perhaps this unconscious bias can be reduced in real life if we know that a woman has objective credentials to back up her expertise? Then at least our initial expectations will be higher. But we may still recoil if the woman asserts that expertise by disagreeing with us. And for men, in particular, even proof of a woman's competence may not be enough.

As Linda Carli of Wellesley College, who has spent a lifetime researching influence, writes: 'A woman who behaves in a competent and assertive manner is often less influential, particularly with men, because she lacks legitimacy.' Or is at least deemed by them to lack legitimacy. For some men, Carli claims, a competent woman can threaten their sense of entitlement to power, so they 'are likely to be resisted as leaders or agents of influence'.[6]

Louise Richardson has noticed this a lot. 'I've certainly had some wonderful male colleagues make fun of how uncomfortable many male colleagues are with me in my present role,' she told me. 'They say, "They really hate it!" They've always believed from day one that men are superior and should be top dog, that's what feels natural to them, so when people challenge this natural order, they don't like it. Some of them may feel personally inadequate.'

Even very senior, highly competent women feel as if they're pushing uphill when trying to influence a room full of men. Dame Sara Thornton is now the UK's Independent Anti-Slavery Commissioner, and used to be Chief Constable of Thames Valley Police. So, like Major General Sharon Nesmith, she has spent her working life surrounded by pretty masculine men. She told me what tactics she used when contributing to the Association of Chief Police Officers' Terrorism Committee. 'There were very few women and I was determined to have my voice heard. So I would always make sure I'd read the papers, which many of my colleagues hadn't done. And I would have thought beforehand about what I was going to speak about. And I would always watch what happened. I was really disciplined, I would write three points in the margin, make them and not ramble. I was always trying to manage the process and manage my ability to influence. But I had to really work at it. To begin with, maybe I was allowed to speak, but it took time to get influence. I was pedalling really hard to get my voice heard, to speak and then to influence.'⁷

The problem is that we all – women as well as men – tend to associate leadership, and therefore influence, with men. Ask people to draw a picture of a leader and it will almost always be a man. Centuries of male dominance have imprinted that on our brains. Within the family, fathers often dominate mothers, so children learn to assume from very early on that men are in charge, that their views trump their mothers'. And the world around us entrenches these stereotypes, as we'll see in Chapter 10. As a result, it is much harder for women's leadership potential, and ability to influence a team, to be recognized.

The best way of demonstrating this, as with the job applicants in earlier chapters, is to conduct an experiment in which the only difference is a male or female name. In one study, participants were asked to call in to a meeting of the sales team of a fictional insurance company. They were addressed by an 'Eric' or 'Erica', reading from the same script, and were asked to rate this person on the extent to which they had exhibited leadership, influenced the team or assumed a leadership role. The Erics who came up with constructive ideas for improvement were rated as better leaders than those who only criticized the team's performance. But Ericas weren't rated better, even though they came up with exactly the same ideas.

The same researchers decided to try this out in real life. They asked participants in a big competition at the West Point Military Academy to choose, after the competition, who they wanted to be team leader. Only the men who had spoken up with ideas were chosen. But, as one of the study's authors, Kyle Emich, says: 'It didn't matter whether women spoke up 1) almost never, 2) rarely, 3) sometimes, 4) often, or 5) almost always. Women did not gain status for speaking up, and subsequently were less likely (much less) to be considered leaders.'[8] The idea of a woman as leader was just too incongruous to be entertained. And if a woman did act assertively by speaking up, the others resisted her influence.

Interestingly, this is one of the few areas in which black women are actually treated better than white women. Because assertiveness is expected more in black women, thanks to the stereotypes we have, if they display it, they are not penalized. One study showed people a picture of a fictional senior manager in a *Fortune* 500 company dealing with a subordinate who wasn't performing well. Dominant leaders demanded action and were assertive; communal leaders encouraged the employee and communicated compassionately. Participants rated the leader on questions such as how well the leader handled the situation and how much they thought employees admired the leader.

While people were negative about assertive black men and white women, black women were given as much latitude as white men to be assertive. This suggests that black women face fewer challenges than white women when it comes to perceptions of leadership (though they have a harder time reaching a leadership position in the first place, because of racial bias compounding the gender bias). 'Black women leaders occupy a unique space,' said one of the authors of the study, Ashleigh Shelby Rosette of Duke University. 'These findings show that just because a role is prescribed to women in general doesn't mean that it will be prescribed for black women.'[9]

For non-black women, though, Carli has discovered that men are more likely to be influenced if women use more hesitant, tentative, self-deprecating language than if they are assertive, even though men think these hesitant women are less competent. Men, but not women, in her study, said that highly competent women were more threatening and less likeable than less competent women, and this reduced the influence the competent women wielded on men.

So all the advice that women are given about not hedging their statements, not apologizing, not doing themselves down, not using uptalk may, paradoxically, be going against their own interests. Yet hesitancy and self-deprecation do make them sound less confident and competent. It's one of so many double binds that women find themselves in when they simply want to be treated the same as men. And these double binds make narrowing the authority gap much harder. If it were simply a question of women being as expert and confident as men, we would have solved it by now. But the very act of being expert and confident can make men resist a woman's authority and influence even more.

Another study tested whether men were more likely to be influenced by 'hyperfeminine' women: not ones who look very feminine, but those who believe in traditional gender roles.[10] Sure enough, the men found the 'high hyperfeminine' woman more persuasive (and more physically and sexually attractive, though they had only heard her voice) than the 'low hyperfeminine' one, even though she was reading from the same script. And men were more likely to be persuaded by the high hyperfeminine woman's argument even though they believed that she was less knowledgeable and competent.

This is all deeply depressing for women. The more competent and knowledgeable you are, the less influential and the more dislikeable you are likely to be to men. Instead you have to be attractive, hesitant and submissive to have any chance of winning them over. It feels like we're back in the 1950s, or living in Harry Enfield's brilliant sketch 'Women, Know Your Limits!' (If you haven't watched it on YouTube, give yourself a treat.)

Women are not even allowed to get angry to blast their way through this prejudice. One study brought people into what they thought was a mock jury deliberation about a murder case.[11] In fact, the other five jurors were actors who were playing out a script: four agreed with the participant, but one held out against them. If the holdout, male or female, showed no emotion, the participant's mind wasn't changed. If the holdout was a man and he showed anger, the participants' confidence in their own verdict fell dramatically. But if the holdout was a woman and she showed anger (while saying exactly the same words as the man), the participants' confidence in their own verdict rose significantly. She had no influence at all.

Before you despair entirely, there is one way in which it is possible to be competent and influential to men as a woman. We shouldn't have to act differently and, in an ideal world, we wouldn't. But either we rage about other people's unconscious (or conscious) bias and don't get very far, or we try to find a way past it. Of course I want this bias to be reduced, and I have countless suggestions in the last chapter for helping that to happen, but while it does still exist, there is at least a narrow way through.

For the research suggests that, as with confidence, what women have to do is to temper their competence with warmth. Warmth is something that is expected in women, but not in men. Women are expected to show communality, while men are expected to show agency. These stereotypes are not just descriptive, but prescriptive too, for women at least. If men are communal, against stereotype, they are rewarded for it. But if women are agentic, they are all too often punished. This is what lies at the heart of the authority gap, at least when it comes to leadership. Women aren't expected to act in the agentic way that leaders are expected to behave in, so it is much harder for them to be accepted as leaders. Their authority is likely to be resisted.

A male leader can be described as 'tough', and we can admire him for it. A female leader who is 'tough' is automatically dislikeable because she goes against our stereotypes: the heuristics in our brain that tell us not just what women are like, but what we think they should be like. We might even call her a 'bitch'. We are much more likely to feel uncomfortable if a woman fails to display communality than if a man does. This is particularly true if a woman also has to show agency, which is what leaders have to do.

So it is important for women to display communality if they want men to listen to them. Women who use rapid, unhesitating and clear language, which is associated with competence, are less influential than men who communicate in the same way; but if women combine this with warmth, such as smiling and nodding, they become as persuasive as their male counterparts and more persuasive than women who show only competence. In general, people who are communal are more liked, whether they are male or female. But because being warm and likeable is prescriptive for women, but not for men, likeableness leads to influence for women more than it does for men. In other words, a man can influence others even when they don't particularly like him, but a woman generally has to be likeable if she wants to be influential and there-fore command authority. This is grossly unfair: why should women have to tick boxes that men don't? Unfortunately, though, until we manage to change the ways in which our brains instinctively work – or at least, correct for them, which we should all try to do – women are going to suffer from this handicap.

Women and girls absorb this very early. Charlotte Stern, 20, works as a catering assistant at her local hospital. She and her friends, she told me, 'find ourselves enraged frequently over this authority gap. As young women we have often felt that our young age and gender combined can be a barrier to progress. I often find myself feeling that I shy away from "big" leadership/authoritative roles for fear of be-coming a "bitch" or being "bossy". When I have to confront other people within a team, I change my approach in order to not come across as a "nasty girl". Yet for my male peers, I feel like they don't have this issue, as they seem "in charge" or expressing their authority.'

Izzy Radford, also 20, a TV development assistant, describes the problem with young men of her age: 'Boys don't listen to you talk-ing about feminism or women's rights, but if a man does, he's seen as so woke and amazing and intelligent. You always have to keep so controlled in a conversation about these things to avoid being called "hormonal", "angry", a "feminazi", etc. You just can't win and it's incredibly infuriating and upsetting. I have found that often you have to be the "right sort of feminist", especially with young men: you have to be funny enough or pretty enough to be allowed to stand up for women without being seen as a bitch.'

Personally, I can tell that I've been socialized to overlay authority with warmth in order to reduce other people's instinctive hostility. I smile more than the men around the board table and, if I'm giving a talk, I smile a lot. I take care to ask people about their families and show genuine interest in their children. Although it's irritating and unfair that men don't have to do it, it doesn't bother me too much.

Perhaps because of the way I was socialized, or perhaps because it's in my nature, I prefer to be nice to people, to show concern about their welfare and to give praise where it's due. It's certainly less of a sacrifice than deliberately talking hesitantly and downplaying my expertise or coming over as hyperfeminine and traditional. Neither of those would be authentic. But if warmth is what it takes to have men listen to me, that's not the end of the world.

Most of the women I interviewed for this book have independently come to the same conclusion. They say they use warmth to defuse any hostility to their authority. 'It's important always to demonstrate competence and empathy,' Muriel Bowser told me. 'You also have to be decisive. You have to cultivate a style to do that.' She was the one who had 'BLACK LIVES MATTER' painted in giant yellow letters on 16th Street, across from the White House, just days after the death of George Floyd, and renamed that stretch of road Black Lives Matter Plaza. That was surely a demonstration of both decisiveness and empathy.

Rania Al-Mashat is an economist and politician who, at the time of writing, is Minister for International Co-operation in the Egyptian government. As a senior woman in the Middle East, she is well used to navigating the difficulties of wielding authority in a very masculine world. In Cabinet, she says, 'I don't engage in conflicts. So one of my colleagues said, "You're like Switzerland: you have your armies, but you don't engage in war." And I think that's where the charm part comes in and the emotional intelligence.'[12]

But the demand to display warmth can have its costs. Helle Thorning-Schmidt thinks that it sets the bar higher for women. 'One of the things that's very significant in terms of how you view women as leaders is that you want to feel their passion, their warmth, much more than you do with men,' she told me. 'And I actually think it weakens how women are perceived, because women have to perform on a broader spectrum than men, and that makes it easier to fail.

'I experienced that to a certain degree, because people said, "We can't feel you, we can't feel what kind of person you are," and those demands were just never, ever made of male prime ministers. For a female leader, you want to see that she's making decisions with her heart and with her brain, but that also becomes a weakness because true leaders, of course, traditionally do not make decisions with their heart. All leaders have to be very rational, but when women are very rational, they are accused of being cold. So, that's the unbreakable dilemma that you have as a female leader.'

Rebecca Kukla is a professor of philosophy at Georgetown University. She despairs over how to overcome this double bind, or as she calls it, the multidimensional bind. 'There is almost no correct way for a woman to use her voice and hold her body to project the proper kind of expertise and authority in a conversation . . . If we sound too feminine, sounding feminine in this culture is coded as frivolous and unserious. If we sound too unfeminine, then we sound like we are violating gender norms or like we are unpleasant or trying to be like a man. If we try to be polite and make nice, then we come off as weak. If we don't make nice, then we're held to a higher standard for our appropriate behaviour than men are. I think there's almost no way we can position ourselves so that we sound as experts. So oftentimes, the content of our words matters less than our embodied presentation as a woman.'[13]

If we are warm and friendly, does that mean we can't also be tough and decisive, as Kukla suggests? Baroness (Brenda) Hale doesn't think so. She asked her pupil-master when she was a young trainee barrister why, given that he was married to a professional doctor, he disapproved of female barristers. 'He said, "The Bar is a fighting profession, medicine is a caring profession." His view was that women didn't know how to fight.

'To be a successful barrister,' she told me, 'you do have to be a fighter and you do have to have the judgement to know when to fight and what to fight and how to fight, and when to compromise and settle. So he was right about that, but he was wrong about women.' There is no reason to suppose that women can't fight just as well as men, whether or not they are combining it with warmth.

If people like her pupil-master believe women can be communal but not agentic, there are others who believe that agentic women can't be communal: in other words, clever, confident women can't be nice. Highly competent women are often assumed to be dislikeable even if they aren't, and again this is a problem that afflicts women but not men. As recently as 2018, Natasha Quadlin of Ohio State University sent out more than 2,000 job applications from fictional recent graduates.[14] On the CVs, she put their grade point average, a measure of academic achievement. She found that the highest-achieving women were called back – either invited for interview or contacted for a conversation – less often than even the lowest-achieving men. And the effect was even starker for maths graduates. The most successful women were moderate achievers: those who averaged B rather than A grades.

She found that while men tended to be hired based on their apparent competence and commitment, women were hired based on likeability. And the moderately clever women were thought to be more likeable than the highly intelligent ones. For men, likeability was barely mentioned.

Here's an example of one assessment of a high-achieving young woman's CV. 'Stephanie seems over-confident and very smart. She would be overqualified for any position in my company. Also, she doesn't quite seem socially warm. Not sure why, there's nothing wrong with being confident, but I get the feeling she's arrogant.' These judgements were made on the basis of her CV alone, and she wasn't even offered an interview.

*

What conclusion can we draw from this? We encourage our daughters to work hard at school and university so that they can get good grades and increase their chances of ending up in a great job. But perhaps they would be better off getting just moderate grades – though I would never advocate that as a solution. We encourage young women to take STEM subjects, such as maths, yet those who excel in the subject are penalized even more severely for their talent.

We think we live in a meritocracy, in which hard work and achievement win rewards. We think we live in a more gender-equal society in which employers are falling over themselves to recruit really bright women to improve the diversity in their ranks. Yet here are new female college graduates, nowhere near the age of parenthood, being discriminated against – and assumed to be unlikeable – for having achieved too much.

Women suffer interruptions, challenge and dismissal when they try to get their point across. Their influence is less than men's because people underestimate their expertise and resist their authority. And if they are experts, they can suffer if they challenge the conventional wisdom. Being highly competent can actively reduce their influence with men unless they play it down or make sure to temper it with warmth. And even if they do, they may be penalized if they are exceptionally able.

This is all concrete – and depressing – evidence that the authority gap diminishes women's influence, particularly over men. Women have to work furiously at being likeable if they want men to listen to what they say. But what if men aren't even prepared to expose themselves to women's views in the first place? That's what we're going to look at next.

Hello? Anyone there?
Voices in the void

'I have a terrible confession to make – I have nothing to say about any of the talented women who write today. Out of what is no doubt a fault in me, I do not seem able to read them. Indeed I doubt if there will be a really exciting woman writer until the first whore becomes a call girl and tells her tale.'

– Norman Mailer

WHEN I WAS a political columnist, I would often be asked to appear on *The Daily Politics*, a BBC TV programme for political aficionados presented by Andrew Neil. On Fridays, they would invite two journalists to sit with Neil for the entire hour and have a discussion after each of the six items. One Friday, I was delighted to be asked on with my former *Times* colleague Daniel Finkelstein.

Danny and I used to sit next to each other at work. We had lunch in the canteen several times a week and would always chew the political fat together. We saw each other as absolute equals, though he was better connected with the Conservative Party and I with Labour. There was no authority gap between us, and he is one of those delightful men who are perfectly content to acknowledge when a woman knows more about something than they do and are happy to cede her the floor.

So imagine my annoyance when we appeared together on the programme and the presenter seemed almost to ignore me. After it was over, feeling short-changed, I decided to watch it back. Sure enough, Neil went to Danny first on all six occasions, and often came back to Danny for a follow-up after I had made a point but never came back to me.

You would have expected him to alternate between us. I'd have forgiven him if the ratio had been four to two in Danny's favour. But six to zero? It just felt so disrespectful of my authority and expertise.

My response was to write a column using this incident as one example of men rating each other more highly in political journalism. It's interesting that, when I commissioned the big data company Lissted to do a survey of following and retweeting behaviour on Twitter across a range of influencers in the Westminster world, the results showed that many leading male influencers followed a highly disproportionate number of male political journalists over women, even after the figures were corrected for the fact that there are more male political journalists than female ones. The survey showed that Neil, for example, was following three times more men than women.

Why might following behaviour matter? It helps to show how much (or little) people want to listen to what others say. We tend to choose people to follow on Twitter because we rate them and think that they will have interesting ideas. If we're not allowing their tweets into our news feed, it implies that we have no interest in their views.

Of course, I don't know why Neil followed many more men than women. But interestingly the gap wouldn't be thought unusual in America. A study of Washington political reporters found that, even though women make up nearly half of the press corps, the male Beltway journalists engaged almost entirely with other men.[1] Of the twenty-five reporters who received the most replies from male political reporters, *none* were women. The male journalists replied to other men 92 per cent of the time. And they retweeted other men 75 per cent of the time. 'I've never seen statistical significance like this before,' said lead author Nikki Usher, who conducted the study while a professor at George Washington University. How can a professional woman expect to get herself heard in that frosty climate?

Most of the bias I've written about is present in both women and men, and we'll look at women's bias in more detail in the next chapter. All of us – to a greater or lesser extent, and often without realizing it – tend to expect less of women, listen to them less attentively and feel uncomfortable with them in positions of authority. But this chapter is going to explore a particular phenomenon: men (like Norman Mailer) not even exposing themselves to women's voices in the first place, across the cultural spectrum, whether the women are on social media, writing books or appearing in films. If these men are not listening to, reading or watching women, how can they accord them any authority at all? How do they even know if the women are any good?

The easiest way to measure this phenomenon – of women whistling into the void – is to look at the books that men and women read. Non-fiction books are sources of authority on a subject; fiction takes us into other humans' worlds and minds, broadening our empathy and understanding. Before going any further, may I ask you to take a moment to think about the last five or ten books you've read, and count how many are by male and how many by female authors? If you're a man and your tally is roughly 50:50, congratulations, you are very unusual. As the writer Grace Paley once said, 'Women have always done men the favour of reading their work and men have not returned the favour.'[2]

The first study to look at this was a little anecdotal, but still telling. Lisa Jardine and Annie Watkins of Queen Mary University interviewed a hundred academics, critics and writers about their fiction reading habits.[3] Four out of five of the men they talked to said the last novel they had read was written by a man, whereas women were almost as likely to have read a novel by a male author as a female one. When asked which novel by a woman they had read most recently, a majority of men found it hard to recall or could not answer. When asked to name the 'most important' novel by a woman written in the last two years, many men admitted defeat and confessed they had no idea. Female authors make up as much of the modern literary canon as male ones, so these men were reading only half the canon, while the women were sampling it all. Maybe the men assumed that novels by women weren't as good, but how could they tell if they weren't even reading them?

As the report concluded: 'Men who read fiction tend to read fiction by men, while women read fiction by both women and men. Consequently, fiction by women remains "special interest", while fiction by men still sets the standard for quality, narrative and style.' If you think about the 'great American novel', I bet you immediately associate it with Mark Twain, John Steinbeck, Philip Roth or Jonathan Franzen. But what about Toni Morrison, anyone? Harper Lee? Alice Walker? Donna Tartt?

The Irish novelist John Boyne remembers attending a literary festival where three established male novelists were referred to in the programme as 'giants of world literature', while a panel of female writers of equal stature were described as 'wonderful storytellers'.[4] He actually believes that women are better novelists than men because, he claims, they have a better grasp of human complexity. 'My female friends, for example, seem to have a pretty good idea of what's going on in men's heads most of the time. My male friends, on the other hand, haven't got a clue what's going on in women's.'

Boyne is unusual, though. Most people seem to have different expectations and therefore different standards for men's and women's fiction writing (as we saw in Chapter 1, when Catherine Nichols sent out her manuscript under a male name). This is what allows some men to believe that women's writing is not worth reading. The Irish novelist Anne Enright explained this beautifully in the *London Review of Books*: 'If a man writes "The cat sat on the mat" we admire the economy of his prose; if a woman does, we find it banal. If a man writes "The cat sat on the mat" we are taken by the simplicity of his sentence structure, its toughness and precision. We understand the connection between "cat" and "mat", sense the grace of the animal, admire the way the percussive monosyllables sharpen the geometrics of the mat beneath. This is just a very truthful, very real sentence (look at those nouns!) containing both masculine "mat" and feminine "cat". It somehow Says It All. If, on the other hand, a woman writes "The cat sat on the mat" her concerns are clearly domestic, and sort of limiting. Time to go below the comments line and make jokes about pussy . . . '[5]

Mary Beard told me how she was once on a book prize panel, 'and it was absolutely clear to me that the men picked really lengthy books. They would pick them up and say, "This is a really weighty contribution" and what they meant was, "This is a very male contribution." Then one of the other judges said in the end, "We're going to have some short books." It's not that the contributions of men and women are colossally different, but the affirmative adjectives that are used to make us think that we can all agree about this candidate rather than the other one tend to be heavily correlated with male candidates. Women don't do "weighty" things. These are words that are not so glaringly correlated with gender, but they're a code for gender. The men who use those words are as unaware as anybody of that.'[6] We all know instinctively that 'heavyweight' is code for 'male'.

The novelist Kamila Shamsie has sat on a number of prize judging panels. 'There are male judges and female judges,' she told me, 'and the women judges are putting forward books by both men and women that they think should be shortlisted. And the male judges are largely putting forward books by other men.'[7]

One year, she decided to call it out. As the panel sat down for their first meeting, Shamsie drew attention to the fact that the prize had only ever been won by one or two women. The next time they met for the longlisting, each of them having supposedly read all the books submitted for the prize, a male judge had on his longlist a couple of books by female authors. When he was asked about one of these books, he shrugged, looked at Shamsie, and said, 'Well, I put it on because such a noise was made on the first day about women not being included.' He clearly hadn't read it, she says.

'But then the really telling thing,' she goes on, 'was it got on the longlist because others of us, which is to say the women judges, had read it and really liked it. And when we came in for the next meeting, this male judge was a very passionate advocate of this book, which he had finally read and found to be wonderful!'

Amanda Craig is a British novelist who writes state-of-the-nation fiction. Yet, as Sarah Hughes asked in the *Independent* when reviewing Craig's latest book, *The Golden Rule*, 'Why is Amanda Craig not better known? Her novels, which have tackled topics from the plight of undocumented migrants to the effects of Brexit on rural Britain, fit neatly into the sphere occupied by the celebrated likes of Jonathan Coe – yet somehow she is not a household name.[8]

'This is possibly because the influences Craig draws on are seen as "female" and lacking in weight. She has never been afraid to reference myth and fairy tale in her work, previously playing on the story of Theseus and Ariadne and spinning comic gold from an updated retelling of *A Midsummer Night's Dream*. Yet it seems that an interest in how stories feed the imagination is considered less compelling than tricks of language or form.'

And very similar books written by men and women can be judged differently. Kamila Shamsie won the 2018 Women's Prize for Fiction for her novel *Home Fire*. A retelling of *Antigone* in the context of the war on terror, it covers deep contemporary themes with an overlay of complex relationships between three Anglo-Pakistani siblings and the son of the Home Secretary. But, she told me, 'When my books get talked about, people go much more to the familial and the romantic elements of them. And actually, the men are writing as much about romance and family, maybe more, but they get talked about in terms of the larger political stories that they're telling.'[9]

Is this just a phenomenon of the English-speaking world? Definitely not. The Norwegian author Karl Ove Knausgaard was feted for his six-volume autobiographical novel *My Struggle*, a minutely detailed account of his domestic life that would probably have been deemed inconsequential if written by a woman. In 2010, Belgian author Bernhard Dewulf won a prestigious Dutch literary award, the Libris Literatuur Prijs, for a semi-autobiographical account of day-to-day life with his children. Three years earlier, a jury of the same prize lamented the fact that women so often write about 'personal trifles'.[10]

To see a better and bigger picture of the extent to which men were failing to read books by women, I asked Nielsen Book Research, the gurus of the book trade, to reveal definitively who exactly was reading what. I wanted to know not just whether female authors were deemed less authoritative than men (possibly because they are judged by double standards), but whether they were even being read in the first place. And the results bore out my suspicion that men were disproportionately unlikely even to open a book by a woman. Overall, looking at the top-selling books (fiction and non-fiction) in the UK, women read slightly more than men: the readers of these books were 54 per cent female and 46 per cent male. But when you break them down by author, the results are dramatically different.

For the top ten bestselling female authors (who include Jane Austen and Margaret Atwood, as well as Danielle Steel and Jojo Moyes), only 19 per cent of their readers are men and 81 per cent women. But for the top ten bestselling male authors (who include Charles Dickens and J. R. R. Tolkien, as well as Lee Child and Stephen King), the split is much more even: 55 per cent men and 45 per cent women. In other words, women are prepared to read books by men, but many fewer men are prepared to read books by women. And the female author in the top ten who had the biggest male readership – the thriller-writer L. J. Ross – uses her initials, so it's possible that her male readers weren't aware of her gender. What does that tell us about how reluctant we are to accord equal authority – intellectual, artistic, cultural – to women and men?

Margaret Atwood, self-evidently a writer who should be on the bookshelves of anyone who cares about literary fiction, has a readership which is only 21 per cent male. Male fellow Booker Prize winners Julian Barnes and Yann Martel have nearly twice as many (39 and 40 per cent). Hilary Mantel has only 34 per cent male readers.

It's not as if women are less adept at writing literary fiction. Quite the contrary. In 2017, all five of the top five bestselling literary novels in the UK and Ireland were by women, and nine of the top ten.[11] And it's not as if men don't enjoy reading books by women when they do read them; in fact, they marginally prefer them. The average rating men give to books by women on Goodreads is 3.9 out of 5; for books by men, it's 3.8.[12]

Turning to non-fiction, which is read by slightly more men than women, the pattern is similar, though not quite so striking. Men still read male authors much more than female ones, but the discrepancy isn't so large because women tend to do the same in favour of female authors. But there is still quite a difference. Women are 65 per cent more likely to read a non-fiction book by the opposite sex than men are. And that suggests that men, consciously or unconsciously, do not accord female authors as much authority as male ones. Or they make the lazy assumption that women's books aren't for them without trying them out to see whether this is true.

This is not just bad for women's book sales. It narrows men's experiences of the world. 'I've known this for a very long time, that men just aren't interested in reading our literature,' Bernardine Evaristo told me. 'So what does that say about our society? Our literature is one of the ways in which we explore narrative, we explore our ideas, we develop our intellect, our imagination. If we're writing women's stories, we're talking about the experiences of women. We also talk about male experiences from a female perspective. And so if they're not interested in that, I think that it says a lot and it's very damning and it's extremely worrying. It seems to me that we're seen as less important and more insignificant. And that is a big problem.'[13]

If women writers do want to be feted, it helps if they write mainly about men. Nicola Griffith analysed the six most important US and UK literary fiction awards over fifteen years and found that the more prestigious the award, the more likely the novel would have male main characters.[14] So, for instance, in fifteen years of the Pulitzer Prize, between 2000 and 2015, more than half the winners were books by men, about men. Of the female winners, half the novels were about men and half were about both men and women. No winner, either male or female, wrote a book starring women or girls. Yet novels are supposed to be about the human condition, not just the male condition. 'Either this means that women writers are self-censoring, or those who judge literary worthiness find women frightening, distasteful, or boring. Certainly, the results argue for women's perspectives being considered uninteresting or unworthy,' writes Griffith.

This lack of acknowledgement is also very hurtful to female authors. Dolly Alderton is a highly successful writer, whose memoir *Everything I Know About Love* was a *Sunday Times* bestseller and won the 2018 National Book Award for best autobiography. Yet in Britain, at least, it had almost no interest from men. Every newspaper and magazine journalist sent to interview her was a woman and it was, as she told me, 'marketed and perceived and received as something incredibly niche by dint of my gender. Yet a female experience is not a niche experience; it's a universal common interest.'[15]

This has really dented her morale. 'I feel like I have no male readers. There's something innately very patronizing about knowing that half the population considers my thoughts on anything to be completely irrelevant to them. I do find that quite upsetting sometimes. On low days when I think about what that dismissal of my thoughts and stories and work is, it's wounding. It sends you into a weird existential place to think that half the population isn't interested in what you've got to say.' It's not just that there is an authority gap: there is a complete void if men aren't reading books by women.

Yet, when she went on a publicity tour to Denmark, a rather more progressive country, it was quite different. She told the male journalist who had been sent to interview her that he was the first ever. 'He couldn't believe how weird that was. He was in his twenties and said he and his friends read memoirs or fiction by women just as much as those by men.' Things *can* be different. And it's a very easy problem for men to fix. All they have to do is actively seek out books by women.

But the UK and the US still have a very long way to go. When *Esquire* magazine drew up a list of 'The 80 Best Books Every Man Should Read'[16], described as 'the greatest works of literature ever published', only one was written by a woman, Flannery O'Connor, and she had a gender-neutral first name. Female authors from George Eliot to the Brontë sisters to J. K. Rowling have had to change or disguise their names to persuade men and boys to read them. I was very tempted to publish this book under the name of M. A. Sieghart.

You might expect *Esquire*'s list to be a matter of blokes recommending to blokes books written by other blokes. But would you expect it from, say, the *New York Review of Books*? In 2019, only 29.6 per cent of its reviewers were women.[17] (The *London Review of Books*, at 32 per cent, was only marginally better.) And of the books it reviewed, only 31 per cent were *by* women. So, just as men are reluctant to read female authors, they are also reluctant to review or recommend them. Even in the *NYRB*, we have blokes recommending books by other blokes. The cultural gatekeepers, the people given authority to pass judgement on books, the critics, are mainly men. And they are according authority to books written by other men. How can women writers expect to be taken seriously by men if male critics largely ignore their existence?

'Affinity is a joyful thing,' writes the novelist Anne Enright. 'I have often admired the ease with which men praise books by other men, and envied, slightly, the way they sometimes got admired in their turn. This spiral of male affection twists up through our cultural life, lifting male confidence and reputation as it goes. Work by men is also read and discussed by female critics; only one side of the equation is weak: the lack of engagement with women's work by men.'[18]

It's not as if the *New York Review of Books* is doing this unknowingly. The VIDA Count, from which I took these statistics, has been publicizing its valuable work since 2010. The percentage of female reviewers in the *NYRB* has crept up since then, but only at the pace of a remarkably idle sloth. It is only three percentage points higher than it was in 2014.

Some publications now do much better. The *New York Times Book Review*'s book pages had 58 per cent women reviewers in 2019, a huge improvement over the decade. The *Times Literary Supplement* had 49 per cent (but it still reviewed twice as many books by men as by women). Again, things *can* change.

Whole genres are often dismissed by the book review pages. The easy-reading thrillers and crime mysteries and speculative fiction that men enjoy are reviewed. No new book by Lee Child goes unnoticed. And when male authors such as Nick Hornby or David Nicholls write commercial books that deal with relationships and family life, they too get reviewed. But the female equivalents are often categorized derisively as chick lit, women's fiction or romance and are usually overlooked by serious papers.

'If chick-lit novels are reviewed, it's to be rude about them,' says Serena Mackesy.[19] She started writing fiction in the 1990s, soon after the first Bridget Jones book was published. So her publishers decided to package her as a chick-lit author. 'All my books were discussing quite big issues, but the blurbs were, "So and so has a job but can she find a boyfriend?"' Mackesy had no say in how her books were marketed, until she had a brainwave. She decided to change her name to the gender-neutral Alex Marwood and start again. 'I know stacks and stacks of very nice men who aren't at all prejudiced against women, but they just automatically think they won't have anything in common with a book by "Serena".' Her first thriller as "Alex", *The Wicked Girls*, was widely reviewed, praised by Stephen King as one of his top ten books of the year, and won the prestigious Edgar Award for Best Paperback Original. She hasn't looked back.

'It's very clear from my Amazon reviews that my readership is much more male. And I'm about 50:50 in my following on social media, whereas before it was entirely women and the men that I knew. It changed everything for me having a new name.' Her experience was just like that of Catherine Nichols, or indeed of the trans men and the job applicants in earlier chapters. One one-star review on Amazon, though, said, 'I do not like thrillers from female authors. I would never have bought this book if I had known that Alex Marwood was a made-up name for a British female writer.' She has it up on her bathroom wall.

At publishing parties, Mackesy enjoys arriving as Serena and then changing to Alex halfway through. 'People who can barely bring themselves to shake my hand suddenly start fawning in recognition.'

Her conclusion? 'It was amazing to find that I was successful and it's delightful to be talked to with respect by strangers who don't realize you're a woman on social media.'

She is still irritated, though, that male authors such as David

Nicholls, who wrote the highly successful and very readable *One Day*, about an on-off romance between a young man and a woman, are widely reviewed and celebrated. '*One Day* is an excellent novel, but a lot of female authors were surprised to find it praised for elements that are dismissed as "classic chick lit" in their own work.' After all, its subject matter is romance, sensitivity and relationships.

Reviews can, of course, be bad, but recommendations are universally good. So I counted the 'Books of the Year' recommendations in the *TLS*, the *Guardian*, the *Spectator* and the *New Statesman*. Only the *Guardian* asked more women than men to suggest their books of the year. The other titles ranged from 61–70 per cent men, with the *Spectator* the most male-skewed.

The male reviewers at the *Spectator* were four times more likely to recommend books by other men than books by women. The (many fewer) female reviewers were much more even-handed, recommending 42 per cent men, 58 per cent women. At the *TLS* too, the men recommended books that were 69 per cent by men and only 31 per cent by women; the female split was 44:56. Even in the *New Statesman* (where two thirds of the reviewers were male, despite its progressive ethos), the men showed exactly the same 69:31 bias, but it was at least counteracted by female reviewers recommending many more books by women. So male critics are according much more authority to male writers than female ones. And the critics themselves are supposedly authorities on the world of literature. As a result, readers are led to believe, first, that men have more authority in recommending which books to read and, second, that books by men are better than books by women. Neither is true, but both serve to exacerbate the authority gap.

Bernardine Evaristo says she makes a point of trying to redress the balance. 'When I'm asked to recommend books, I nearly always choose women. And I nearly always choose black women or women of colour, because I know if I don't, they're very unlikely to be on those lists. So I make a point of doing it because then at least there are going to be two or three books there that are not by white people and not by white men.'[20]

David Bamman, an assistant professor at University of California, Berkeley, did a similar study in 2018 of the hundred most recent interviews in the *New York Times*'s 'By the Book' column, looking at which books authors said they had on their bedside tables.[21] He found an even stronger male bias. Male interviewees recommended four books by men for every one book by a woman, whereas women were extremely even-handed, recommending 51 per cent men and 49 per cent women. Half the male interviewees mentioned no women writers at all.

This connects across, he says, to the lowly status that women have in novels written by men. 'Men remain – on average, as a group – remarkably resistant to giving women more than a third of the character-space in their stories.'[22]

The author Lauren Groff tried to redress the balance when she did her 'By the Book' interview in May 2018, by naming only female authors. She signed off by asking, 'When male writers list books they love or have been influenced by – as in this very column, week after week – why does it almost always seem as though they have only read one or two women in their lives? It can't be because men are inherently better writers than their female counterparts . . . And it isn't because male writers are bad people. We know they're not bad people. In fact, we love them. We love them because we have *read* them. Something invisible and pernicious seems to be preventing even good literary men from either reaching for books with women's names on the spines, or from summoning women's books to mind when asked to list their influences. I wonder what such a thing could possibly be.'[23] It is surely (some) men's blind spot when it comes to letting women's views into their lives, admiring their writing, and according them literary authority.

This has real-life consequences. As well as not being taken as seriously as men, women writers have to put up with earning less because so many fewer men are reading their books and so their publishers value them less. Sociologist Dana Beth Weinberg and mathematician Adam Kapelner of Queens College-CUNY looked at 2 million titles published in North America between 2002 and 2012.[24] They found that, on average, books by women were priced 45 per cent lower than books by men.

But it has consequences too for us all. 'Women's voices are not being heard. Women are more than half our culture. If half the adults in our culture have no voice, half the world's experience is not being attended to, learned from, or built upon. Humanity is only half what we could be,' writes Nicola Griffith.

Which is why it was so cheering that the Man Booker Prize, whose previous record was only marginally better than the Pulitzer, decided to award its 2019 prize jointly to Margaret Atwood's *The Testaments* and Bernardine Evaristo's *Girl, Woman, Other* – both novels by women, about women and, in Evaristo's case, about women of colour. Progress is finally being made; let's hope that it spreads more widely. Let's hope that women's writing starts to be valued as much as men's and that men start to open their minds to the other half of the world.

It wasn't until she was sixty that Evaristo felt her work was at last being taken seriously. 'It's always been a struggle to reach a wider audience, and to be conferred with the kind of respect that white men have traditionally had in our literary culture in the UK,' she told me.[25] And to what extent was this because she was female or black? 'Sometimes we have to break it down, don't we? Are we talking about race? Are we talking about gender, are we talking about class, are we talking about education? All those things play a part. The fact that I was the first black woman to win the Booker, or the first black British person to win the Booker, tells a story in itself.'

And how did it eventually happen? 'Because of who was in the room. I think it's that simple. There were four women on the panel, and one man, and I think that that made history. They were four very strong women, and my book clearly spoke to them. At the same time our society has changed somewhat in the last few years, in that there is more receptivity to black women's art, ideas, culture, literature. So I think the time was right for a book such as mine to break through. But it did take fifty years for that to happen. And I did have to share the prize, although, as I always say, I will take it any way it comes.'

When I was a child, my brother was constantly badgering me to watch Westerns and thrillers on TV. I sat through loads with him, despite protesting that I didn't like them. He told me I was wrong: these were, objectively, great films.

Because I was young, and took my older brother at his word, I thought the problem lay with me. Eventually, though, I realized what it really was. In Westerns, there were no interesting female characters, just the occasional moll at the bar. And thrillers usually had a plot in which an attractive young woman was stalked and killed by a psychopath. Neither of these bothered him, naturally, because he was a boy.

But these were 'great' films. They had won awards. They were part of the canon. Who was I to say that *The Good, the Bad and the Ugly*, a movie with three main male characters and a supporting cast of fifty-three, of whom only one was a woman, wasn't great?

Yet can you imagine men flocking to see a film which had a cast of fifty-six characters, all but one of whom were female? *Little Women*, directed by Greta Gerwig and released in 2019, was a superb adaptation of the book, and had four strong male characters alongside the female ones, as well as more than thirty men in the supporting cast. Yet still, as *Vanity Fair* put it, '*Little Women* Has a Little Man

Problem'.[26] The screenings organized by Sony Pictures for awards season voters attracted disproportionately few men.

'I don't think that [men] came to the screenings in droves, let me put it that way,' said the film's producer Amy Pascal. 'And I'm not sure when they got their [screener] DVDs that they watched them. It's a completely unconscious bias. I don't think it's anything like a malicious rejection,' she concluded. But even with this charitable interpretation, the results were the same. Critics said it was fabulous: their reviews gave it 94 per cent on Rotten Tomatoes, five points higher than the acclaimed but all-male *1917*. But while *1917* won Best Picture at both the Golden Globes and the BAFTAs, *Little Women* didn't win Best Picture at any of the major awards – probably because a lot of the eligible male voters didn't even watch it, let alone vote for it.

The male actor who plays Mr Dashwood in the film, Tracey Letts, put it more bluntly. 'I just can't believe we're still having this fucking discussion where movies by men, and about men, and for men are considered default movies. And women's movies fall into this separate and unequal category. It's absurd.'

Just like in the book world, critics in the film industry are predominantly male, outnumbering women by more than three to one in the US.[27] Again, male film critics are more likely to review films with male protagonists, and 51 per cent of reviews written by women but only 37 per cent of reviews written by men are about films featuring at least one female protagonist.

And men are less likely than women to review films with female leads favourably. Women critics award an average rating of 74 per cent and men an average rating of 62 per cent to films with female protagonists. When it comes to male leads, though, there's little difference. Women give them 73 per cent on average, and men, 70 per cent. So it looks as if male critics are less likely to watch movies about women in the first place and, if they do, they are more likely to mark them down. Women are more even-handed.

It won't surprise you, then, to find the same happening in the arts world in general. When researchers looked at elite newspaper coverage of the arts and culture over fifty years in France, Germany, the Netherlands and the US, they found that, despite women's increased involvement in the arts over that period, coverage of their work remained steady at only 20–25 per cent in all four countries over half a century.[28] The title of their article says it all: 'These Critics (Still) Don't Write Enough about Women Artists'. And which gender are the critics? Mostly male. So, again, women's artistic work is either undervalued or ignored, perpetuating the stereotype that women are inferior or unworthy, less deserving of our respect.

Art by women is literally undervalued. On average, it sells for 48 per cent less at auction than art by men. So maybe female artists aren't as good? Well, one experiment showed people art generated by the DeepArt algorithm and randomly assigned male and female names to it. Affluent men who frequently visited galleries – your average art collector, in other words – rated the works by 'men' higher than the works by 'women'. As the authors write, 'Women's art appears to sell for less because it is made by women.'[29] It is not worse; it is just female.

None of this is necessarily deliberate. If boys are brought up to believe, deep down, that they are superior to girls, and if society is delivering the same message to them once they become men, they may well believe that male writers, film-makers, artists, and so on, are more worthy of their attention. And if they see the default human condition as male, they may well perceive men's stories as universal, but women's stories as niche.

But these assumptions must be questioned, because they are based on a false premise: that men are better than women. If we accept that women are as good as men in almost every field that doesn't require physical strength, then we have to start treating them that way, recognizing their talent and according them the respect they have earned. We have to start following them on Twitter, reading their books, watching their films, appreciating their art. We may even be surprised and delighted by how expert they are.

If men are sceptical that women will write about subjects that interest them, they could try Pat Barker on the First World War or Hilary Mantel on the machinations of Henry VIII's court. Novels by women are not all about sappy romance. Men's inattentiveness to works by women, which contributes hugely to the authority gap, can be solved incredibly easily. All these men have to do is actively decide to expose themselves to women's voices. Once they become used to it, they may even find that these turn into human stories rather than niche female ones.

Men can gain a huge amount from broadening their minds and their tastes. Just because a book is written by a woman or a film is about women doesn't mean it has nothing to offer them. It opens their eyes to what it's like to live as a woman in the world, the first step to learning empathy. And it may help to burst the bubble many men have been inadvertently living in and allow new thoughts and insights and ideas to germinate in their heads. Isn't that what the arts are for?

9

Women do it too
How our reptilian brains work against us

'WHEN I'M IN CHARGE OF EVERYTHING, I
WON'T BE RUTHLESS AND PUSHY LIKE SHE IS!'

'It was the single most important and transformative
day of my life, when I came face to face with my own
bias, with the fact that my mind and my hands were
unable to co-ordinate and associate "female" with
leadership as well as "male" with leadership.'

– Mahzarin Banaji, psychologist

W HEN ANNE HATHAWAY played the lead female character in the film adaptation of David Nicholls' *One
Day*, she found herself, unusually, being directed by a
woman. And immediately her own unconscious bias against other
women started raging. 'I really regret not trusting the director of
One Day, Lona Sherfig, more easily,' she admitted to her interviewer
on *Popcorn with Peter Travers*. 'And to this day, I'm scared that the reason I didn't trust her the way I trust some of the other directors I've
worked with is because she's a woman. I'm so scared that I treated
her with internalized misogyny. When I've seen the first film directed
by a woman, I have in the past focused on what was wrong with it
and when I see a first film directed by a man, I focus on what's right
with it. I focus on where he could go with the next one and I focus
on where she failed to go.'[1] As she admitted, she was contributing to
the authority gap by underestimating the ability of other women.

When women are biased against other women, we often call it
internalized misogyny. We have internalized this bias because of
the way we have been brought up, because of what we see around
us in society, and because of the prevailing attitudes of the more
powerful gender. Women are as susceptible to creating stereotypes
as men are, and these form the heuristics that lead our brains to
take short cuts: to judge people by their gender rather than their
individual ability.

Philippa Perry is a psychotherapist and writer. She described to
me how toxic the misogyny of her upbringing was. 'My father definitely believed that women were inferior. He wanted sons and my

mum was forty when they got married. They had two children and I was the second, and I was the biggest disappointment because there wasn't going to be another chance after me and I was a girl. He made jokes about it the whole time, saying, "Even the dog's a bitch." Also, he had a lot of misogyny that I just absorbed. He said things like, "Women are disgusting." It just seeps in. I looked at the TV I was watching in the seventies, and the men were allowed to have characters, but the women weren't. We could be flibbertigibbets, we could be battleaxes, we could be sexpots, but we couldn't be normal people. All this was coming at me under the radar all the time. We were brainwashed into thinking we were "less than".'[2]

Even if, consciously, we believe strongly in gender equality, we may sometimes find ourselves reacting negatively towards other women because of this unconscious bias. We may still find it harder to associate women with careers or men with family, even if we are ourselves professional women. In tests for unconscious – or implicit – bias, slightly more women than men show this bias (80 per cent compared with 75 per cent). The 'gender-career' Implicit Association Test (IAT) shows names such as Rebecca and Daniel, along with home and work words such as 'children' and 'office'. How quickly and accurately you click on particular combinations is supposed to show how strong your implicit associations are between women and home, and men and work.

The Implicit Association Test has its critics. You can take it twice and it doesn't always give you the same result. And just because you find it easier to associate a woman with cooking and a man with an office – the sort of images it measures – it doesn't mean that you believe these stereotypes are a good thing or should be encouraged to continue. But the IAT does at least show us how strongly the stereotypes are embedded in our brain, whether we like them or not.

Depressingly, given that I've written a book about this very issue, my results show that I'm moderately biased. And I'm not alone.

Most of us are slower to associate, say Michelle with 'management' or Paul with 'children' because, as Harvard professor Mahzarin Banaji, who developed the test, told me: 'A thumbprint of our culture has been left on our brain.'[3] We are so used to associating men with taking charge and women with taking care that we find it more difficult to acknowledge the opposite. We are used to seeing men as leaders and women as mothers or subordinates. There is no value judgement involved: it doesn't mean that we believe women *ought* to be in the kitchen. It's just that we find the pairings that go against stereotype rather more incongruous and harder to imagine, which slows us down. Women, apparently, are particularly slow when trying to associate men with 'home' words . . .

If you are asked to sort a pack of cards by putting hearts and diamonds on the left and clubs and spades on the right, you can do it very quickly and easily, simply by dividing the reds and the blacks. But if you are asked to put, say, hearts and clubs on the left and diamonds and spades on the right, you slow right down. This is what happens with the IAT: associating a woman with cooking is instant; putting her at the board table, or her husband at the ironing board, takes a fraction longer.

'It was the single most important and transformative day of my life when I came face to face with my own bias,' Professor Banaji told me. 'Actually coming face to face with the fact that my mind and my hands were unable to associate female with leadership as well as male with leadership.'

Professor Banaji told me a popular riddle and her response to it. 'A father and a son are in a car accident. The father dies. The son is taken to a hospital. The attending surgeon says, "I can't operate on this boy. He is my son." You ask people, how is this possible? And like me, people go through enormously creative incorrect paths to come up with solutions, like mine, which is the father who died was the adoptive father and the father who was the surgeon could have been the biological father, when a very simple answer was staring me in the face, which was that the surgeon was the boy's mother.'

People are just as likely to give the answer that the boy had two gay fathers, even though the likelihood of that being the case is so much smaller than that his mother was a surgeon. And, bizarrely, even personal experience doesn't help. Professor Banaji says, 'I've had a number of people who've said to me, "I was really shocked because my mother was a surgeon and I couldn't come up with the right answer!" That, to me, is a great example of the power of the culture, that it can set aside your own deep personal experience. Your mother being a surgeon is not going to protect you against gender bias, because what you see in the world is driving that.'

When I made a BBC Radio 4 programme about implicit bias, I asked listeners to imagine a hijacker bursting into the cockpit of a plane and attacking the pilot. Then I asked them whether, in their imagination, the pilot was male or female – probably a white man, I surmised. Margaret Oakes tweeted afterwards: 'Driving home in uniform, this programme asked me to imagine a pilot. Yes, I pictured a white male, in spite of being a female pilot!'

Kristen Pressner is Global Head of Human Resources at Roche Diagnostics, a medical tech company. She is honest enough to concede, 'I have a bias against female leaders. No one could be more surprised about this than me. I'm a woman, a leader and, on top of that, I work in human resources, so it's kind of my job to be unbiased. In fact, I passionately encourage women to step into leadership.

'But not long ago, two members of my team asked me to look into their compensation. My first reaction to the man's request was, "Yup, I'll look into it." My first reaction to the woman's request was, "I'm pretty sure you're good." [In other words, you're paid enough already.] I had two very different reactions to basically the same request. I realized that I see men as providers, but not women, which is really interesting because I'm the sole financial provider for my family of six. My husband is a stay-at-home father for our four children. I take charge and he takes care. So I have a bias against women leaders; I have a bias against myself.'[4]

Many (though not all) of the studies that I have cited so far reveal that women show as much unconscious bias as men. Female science professors were just as likely as male ones to prefer the 'male' applicant for the post of lab manager, even though his CV was identical to the 'female' one. Women were just as likely as men to mark down the talkative 'female' CEO but to rate the 'male' one.

In another study, some managers, both male and female, thought that 'male' candidates were more competent and offered them higher pay.[5] What was really interesting, though, was that the managers doing this were the ones who believed that gender bias no longer existed in their profession. Those who believed bias still existed recommended roughly equal pay. Most of the managers who thought gender bias no longer existed were men. But the women who believed this undervalued female staff just as much as the male managers did.

In a different experiment, male and female employers were asked who they would hire for a simple maths task. When the only information they were given was a photograph of the candidate, both women and men were twice as likely to hire a man.[6] This bias did not go away when candidates were first allowed to say how they expected to do on an arithmetic task, mainly because men had a tendency to boast about their performance, while women tended to underestimate their ability. (In fact, both genders performed equally.) Even when the employers were given accurate information about the actual performance of the candidates, the bias did not fully disappear. The more implicitly prejudiced a person was, as measured by the IAT, the less likely they were to correct their bias by taking into account the accurate information.

The IAT shows a slightly stronger bias in women than in men, on average, when it comes to the tendency to associate men with careers and women with family, even when the women themselves have careers. This is mainly explained by men being quicker to associate male names with family words. Only in Austria, Denmark, Finland and Sweden is this not the case. And amazingly, there is very little difference between the young (under twenty) and the old (over forty) in bias against associating women with careers: if anything, the young are very slightly *more* biased: 1.17 to 1.11 on the IAT score[7]. I asked Professor Banaji's co-author, Tessa Charlesworth, whether she was surprised by this.

'Yes!' she replied. 'Honestly, the lack of differences across nearly every demographic group (whether gender, race, age, college, religion, even politics) was very surprising. In social psychology, the expectation has long been that our social identities [such as age] should fundamentally shape our attitudes and stereotypes, especially when they are about other social groups. The finding of *no difference* suggests that the sources of change for gender stereotypes may be particularly widespread, societal movements.'[8] As Professor Banaji says, the predominantly patriarchal culture has left an imprint on all our brains, whatever our age, gender or view of the world.

Our unconscious brain is much more powerful than our conscious one. It likes to organize the world into patterns so that it can react more quickly – in an automatic, reflex way – in order to keep us safe. This process of categorization helps us, in evolutionary terms, to discern friend from foe. As Tinu Cornish, a psychologist at the Equality Challenge Unit, explained it to me: 'Our unconscious brain instantly categorizes people into "Are they like me and are they in a high-status group; or are they not like me or are they in a low-status group?" And then we associate positive characteristics towards people who are like us or who are high status, and negative characteristics towards people who are not like us or are low status. Then comes an association of emotion: warmth towards people who are like us, part of our "in-group", and cold towards people who are not like us or are in our out-group, and it's this process of categorization that drives our behaviour.'[9]

And this, says Cornish, helps to explain how women can be implicitly biased against other women, even though they ought to be part of our in-group. 'Our in-group is not just people who look like us, but also it's the people who we see have high status; and, despite our beliefs, if every time you come to work, every time you switch on the telly or you listen to the radio, men are associated with high status, leadership and competency, that is what our unconscious brain is going to learn.' And if women are portrayed as ditsy shopaholics or cat-fighting bitches, that too is going to feed into our subconscious.

'It can be helpful to think of our unconscious brains as our mammalian or our reptilian brain,' she explains. 'They're not reasoning things out in words, they're learning what things are associated together, and when two events are associated together, our brain actually lays down neurons to connect them. And if our social territory is "think leader, think male," then that's what's going to get reinforced in our unconscious mind despite our dearly held beliefs to the contrary.'

These short cuts or heuristics that our brain takes help to stop it becoming overloaded. We don't have the time or energy to work everything out from scratch every time we make a judgement. But they can also lead us to make the wrong judgements of people. Even if women, on average, are more likely to be stay-at-home parents, it doesn't mean that the one standing in front of you is. Even if more scientists are male than female, it doesn't mean that men make better scientists than women.

The good news, however, is that these implicit stereotypes are starting to weaken over time. For instance, when children were asked in the 1960s and 1970s to draw a scientist, more than 99 per cent of the time they drew a man. Now, the figure is 72 per cent.[10] It shows that changes in how we consciously perceive the world can feed back into our unconscious. If we are prepared to wait for perhaps a century, this will probably be enough in itself to wipe out the authority gap. But although it is encouraging that our unconscious biases are lessening, it is happening very slowly and won't be enough on its own to force a radical change in behaviour in my lifetime.

Professor Banaji and Tessa Charlesworth have analysed two implicit gender stereotypes, one that associates men more than women with science and one that associates men with careers and women with families. They tracked nearly 1.4 million responses to these two IATs from 2007 to 2018, the largest sample ever used to investigate the change in gender stereotypes over time.[11] And what they found, for those of us who hope for greater equality between the sexes, is somewhat heartening.

It looks as if our unconscious biases are slowly starting to dissipate, for all age groups, for men and women, for liberals and conservatives, in virtually every country in the world. Implicit stereotypes fell overall by seven percentage points over that time for both gender/career and for gender/science. In 82 per cent of countries, there was a fall for science, and in 91 per cent of countries, for careers. But the figures are still high: 70 per cent of test-takers still show a gender/career bias and 67 per cent a gender/science bias.

Still, any decrease in bias is welcome. It has concrete results. For instance, Professor Banaji has found a strong correlation between levels of implicit bias in a country against women doing maths and science, and the performance of girls in those countries in maths tests. The lower the bias, the better the girls perform, and vice versa. It's also been shown that girls do better in science tests when they contain illustrations of female rather than male scientists.[12]

Subliminal effects, and other people's expectations, make a difference.

So, as we see more women in the world around us become scientists and more women succeeding in their careers and gaining
authority, our brains should gradually rewire to weaken our unconscious stereotypes and reduce our implicit bias. That positive feedback loop should help to narrow the authority gap further. Women's
ability and opinions are already much more respected than they
were, say, in the 1950s. And that's very welcome. But we still have
some way to go before we can close the gap altogether.

In the meantime, we have to accept the painful truth that women,
too, are biased against women, and this sometimes contributes to
the authority gap at work. 'Women aren't always women's best
friends, professionally,' claims Baroness (Brenda) Hale, after telling
me a story about how she was passed over for a professorship for
which she was eminently qualified because a senior woman at that
university 'felt my application was a bit of a threat to her'.[13] As we'll
see below, there is often room for only one woman at the top, particularly if an organization is run by men.

Hale has now retired as President of the Supreme Court, and
this 'queen bee' syndrome was probably more prevalent in her generation than in the ones that have followed. Queen bees are women
who relish being the only woman at their level and are not inclined
to help others along. If anything, they are more hostile to other
women than to men, seeing them as competition. In the latter decades of the twentieth century, when many women found themselves the only female in a throng of men at work, they sometimes
found that the only way in which they could get on professionally
was to pretend to share their male colleagues' values. If they came
out as a feminist, or argued the case for more women, they ran the
risk of being ostracized.

This was most obvious to me in the 1980s, when I was beginning my career in journalism. Margaret Thatcher was prime minister, and in the eleven years she spent in Downing Street, she appointed only one woman to her Cabinet, who lasted a matter of months. My male colleagues openly disdained the Women in Journalism organization that I had helped to found. But I also had female colleagues who didn't want to be associated with it, and female bosses who clearly resented my attempts to carry on working while I had children. These older women had either had to remain childless in order to get on, or they had to take a career break and sacrifice pay and promotion. They didn't like seeing women from a younger generation having it both ways.

Andrea Jung, the former CEO of Avon, had a similar experience. 'I think that it is a misperception that women are the best and only mentors for other women. I can tell you that I've had women bosses who actually were less supportive than male bosses in the early part of my career. That is not uniform, I've had fantastic mentors who also were women, but I'm just saying it's not always a given and I have witnessed sometimes where women who have not had children, had to make their way to the top one given way, expect all women to give up the same thing or else they shouldn't have the same opportunity.'[14]

If there is such a thing as a Queen Bee syndrome, it may have something to do with the way the different sexes socialize with each other from childhood. Boys and men are more likely to hang out in groups, which can contain quite disparate members, from the most successful/popular to those who are less so. Girls and women, by contrast, are more likely to hang out in pairs, with a best friend of very similar rank. A study by evolutionary biologists at Harvard set out to discover whether this had an effect on the way men and women treated subordinates at work.[15]

They looked at academic publications in psychology from fifty North American universities and found that both male and female professors were equally likely to co-author a paper with another full professor of the same gender. But the male professors were much more likely to write a paper with a male assistant professor than the female professors were with a female one. 'This is consistent,' they claim, 'with a tendency for men to cooperate more than women with same-sex individuals of differing rank.' In other words, senior men were pulling more junior ones up the ladder, but senior women were doing so much less.

Another study, done by Dutch psychologists, looked at how Dutch and Italian professors viewed their Ph.D. students.[16] Although the students had equal publication records and levels of commitment to their work, professors of both genders tended to believe that their female Ph.D. students were less committed to their careers than their male ones. But this bias was held more strongly by women than by men, and most strongly by older women.

For their generation, it was extremely rare for a woman to make it all the way to become a full professor. So perhaps it was something about the context in which older women rose up the ranks that explained their behaviour. They had to fit into a very male environment, they faced more obstacles, more sexism and more outright discrimination than the younger ones. So they had to adapt, almost to show that they were more male than the men, to have a chance of being respected and accepted.

'Do sexist organizational cultures create the Queen Bee?' asked a paper co-written by Professor Naomi Ellemers, one of the authors of the experiment above.[17] This time, the researchers sent out questionnaires to ninety-four Dutch women in management jobs. What they found was that the less women identified with their gender at the beginning of their career and the more discrimination they encountered on the way up, the more likely they were to behave like Queen Bees later. The ones who strongly identified with other women tended to club together to fight the discrimination. Those who didn't tended to set themselves apart from other women, in order to be treated better by men. If it weren't for the sexism in the first place, though, this behaviour wouldn't be necessary.

Professor Ellemers and her colleagues also carried out a study asking Dutch policewomen to recall specific experiences of being discriminated against.[18] They found that being reminded of discrimination prompted the women officers to downplay the sexism they had experienced. It also triggered ostracizing behaviour among policewomen who identified weakly with other women at work.

'They are being taught that to be successful in the organization, you need to adopt male characteristics,' Ellemers says. 'They cope with gender bias by demonstrating they are different from other women.'[19] These women use phrases such as: 'I'm not like the other women, I'm much more ambitious.' Ellemers calls this 'self-group distancing' – a tactic that is also used by other oppressed groups, such as gay men.

Even now, particularly for women in very male-dominated professions, there is a pressure to be one of the lads. And there seems to be a pattern, in these really hard-bitten, masculine environments, of women averting their eyes to sexism and denying that they were treated differently, until they became quite a bit older and recognized that this had been the case. 'If you'd asked me this about ten years ago,' says Major General Sharon Nesmith, 'I probably would have denied that there'd ever been a difference in how seriously I was taken, and I'd say that was largely because I had developed a bit of an immune system, where my strategy for just getting on was not seeing the barriers. It became so normalized to me that I didn't react when I was being treated or viewed in a slightly different way. I've got to a level of personal and professional maturity where I think I can now see it more than I did and, regretfully, I probably should have seen that earlier.'[20]

Janet Yellen faced intense sexism from her male colleagues in her early days as an assistant professor at Harvard. 'It was a very unsupportive environment, and I felt very alone. It was not a collegial place. And co-authorship is something that is very important, people work together. I was not hanging out with the guys.'[21] More importantly, the guys weren't hanging out with her.

It wasn't until three years later, when another woman joined the economics department, that they began to start writing academic papers together. 'That really was an important relationship to me, and I probably never would have gotten tenure at a good university ever, had I not actually done that joint work with this other woman.

'For the next twenty years, if somebody had said, "Do you think the fact you were the only two women in this place had something to do with it?" I probably would have said no. Now, as I look back on it, it seems 100 per cent self-evident to me. But, when I said to this friend, forty years later, "Rachel, do you think we ended up working together because we were the only two women?" – I'd now decided, of course, that was why – she said, "No, I don't think so." And, I thought it interesting that, even after all those years, and all we had learned about the status of women in the economics profession and the troubles that they have, that she would still feel that.' It was as if Rachel had to sublimate any feelings of unfairness in order to get on: like making your way through a blizzard with your head down, your hood up and your scarf wrapped round your face. You have to just put one foot in front of the other, impervious to anything going on around you.

The trouble is that women trying to get on in their careers face both affinity bias, in which men prefer to hire and to promote other men (or co-author papers with them), and gender bias. As Andrea S. Kramer and Alton B. Harris, the authors of *It's Not You, It's the Workplace*, explain: 'Affinity bias and gender bias often work in tandem to make women's same-gender workplace relationships difficult because they limit the number of positions for women at leadership tables, thereby forcing the people vying for those spots into direct competition with one another.

The two forms of bias also create substantial, if not overt, pressure on women to adopt a decidedly masculine management style in order to identify with the male in-group and distance or differentiate themselves from their female peers. These dynamics can foster antagonism between women, which is then often wrongly attributed to their inherent nature, rather than to workplace circumstances.'[22] Senior women can also react by promoting only exceptional junior women, for fear that average ones may just confirm men's stereotypical beliefs that women aren't as good as men. And they don't want to look, to their male colleagues, as if they are being nepotistic by helping more junior women – even though senior men do that constantly to more junior men.

But Kramer and Harris found no evidence that women were more mean-spirited or antagonistic in dealing with other women in the workplace than men were in dealing with other men. It's just that in high-status jobs, people are expected to be decisive, tough and assertive, and when women are like that they are often perceived as cold, hard-hearted and hostile. But what would we call a man who was decisive, tough and assertive? Probably 'alpha male', which has none of the derogatory connotations of 'queen bee'.

The assumption that other women are going to be queen bees can be very damaging. It can prevent women from trusting each other in the workplace and it fuels the sexist stereotype that women love nothing more than bitching at each other and getting into catfights.

Meanwhile, we expect senior women in organizations to do all the heavy lifting of taking charge of diversity programmes and leading women's networks. Many women step up to this job, because they want to help younger women, but it's onerous and undervalued work. If men do this, they are rewarded. If women do, they aren't; and if they refuse to, they are labelled Queen Bees.[23]

The Queen Bees may even be behaving rationally. A recent study of 350 executives showed that men who promoted diversity received slightly better performance ratings. They were seen as the good guys. Yet women who did it were punished with significantly lower ratings. They were seen as nepotistic, trying to give advantage to their own group. So were people of colour.[24]

And these women may also be correct in believing that there isn't room for more than one woman at the top. Companies work hard to appoint a woman – but often only one – to senior positions. Cristian Dezsö and colleagues from the Columbia Business School trawled through the top five jobs at 1,500 businesses over ten years and found that once a company had appointed a woman to a top-tier job, the chances of a second woman getting to the same level dropped by 50 per cent.[25]

The authors had thought that once one woman was appointed, there might be a snowball effect, leading to many more. 'In fact, what we find is exactly the opposite,' Dezsö says. 'Once they had appointed one woman, the men seem to have said, "We have done our job." ' But they found no evidence to support the self-serving argument. If the sole woman were preventing other women from being hired, this would be most obvious in the few companies with a female CEO. But in those cases, a second woman was *more* likely to be appointed to the top five.

In the real world, in fact, it turns out that women are far more supportive of each other than proponents of the Queen Bee syndrome might claim. A study by the Credit Suisse Research Institute of 3,400 of the world's largest companies found that those run by women were 50 per cent more likely to have a female chief financial officer than those run by men and 55 per cent more likely to have women running business units. 'Female CEOs are far more open to and effective in bringing women executives up through the pipeline,' the researchers write. In the UK, in companies in the FTSE 350 that are led by women, on average, a third of the executive committee is female, compared with a fifth in companies led by men.[26]

Thankfully, most of us have personal experience of other women actively supporting us at work. There is often a sisterly camaraderie, a sense of us having to unite to face a common threat. Even at the very top, this still happens. Jadranka Kosor, former Prime Minister of Croatia, told me she had wonderful memories of how kind Angela Merkel was to her when she was fighting for Croatia to join the EU.[27]

Kosor takes up the story. 'Truly great support to me was given by Angela Merkel, who when we met said, very openly, "You do your homework on EU accession and then I'll be able to help you." She also said during our first meeting, "I hope we will be able to become good friends," and I have to say I think this actually happened. She was a big support, especially on one of the most important days of my life.

'On 9 December 2011, during the ceremony of the signing of the accession treaty between Croatia and the EU, there were a lot of speakers, and they would often mention that Croatia had been able to finish the negotiation path thanks to the energy of one woman, Jadranka Kosor. Every time someone mentioned me, Mrs Merkel, who was sitting behind me, would say, "Bravo, Jadranka!" But by then I had lost the election, so I experienced mixed emotions: pride because of this, but also disappointment because of the loss of the election.

'After it was signed, Mrs Merkel came up to me and held out both of her hands and embraced me tightly, and that was when she told me directly, "Do not forget, without you, this would not have happened." I thought if I started to cry now, I would never stop crying! But I made it, and this was the strongest emotion I had felt during a day when emotions were running very high.'

Baroness (Minouche) Shafik, now Director of the London School of Economics, has had a string of incredibly senior jobs: Deputy Governor of the Bank of England, Permanent Secretary at the Department for International Development and Deputy Managing Director of the IMF under Christine Lagarde. 'Everywhere I've worked,' she told me, 'I've had a network of women that I saw regularly and who supported each other. When you're a significant minority in a large organization, it's really comforting to be in a room where you're the majority for a change.

'I can remember once, I was at the IMF, we had a big meeting and there was a break, and we all trooped off to the bathroom. There was me, there was Christine, there were three female directors, and suddenly the women's bathroom was, like, the cool place to be, whereas usually, you always see the men go off and then they share gossip in the bathroom. You think, "What do they know that I don't know now?" It was really nice to have that feeling, of actually, for a change, the women's bathroom was where the power was.'[28]

Not all women are supportive of other women at work. There are still a few queen bees. But they are dying out, and many more women these days try to act in a sisterly fashion to each other. We almost all suffer from unconscious bias – women as well as men – but women are probably more motivated to correct for it. After all, it is they who suffer from it when it is turned against them. So if they find themselves reacting badly and unfairly to another woman, they are more likely to question their bias and put it to one side. And the fact that unconscious bias exists in women, too, doesn't let men off the hook. We all need to civilize and humanize our reptilian brains. We need to notice when we are instinctively judging someone by a stereotype rather than treating them as an individual person. For it is only if we actively notice that we can do something about it.

We also need to work on the outside influences that perpetuate these stereotypes in our unconscious. As we've discovered, our implicit bias is gradually declining, but it is still high and could fall a lot faster if we weren't constantly having it reinforced by what we see in the world around us. That's what we're going to look at next.

It's all around us
The world is framed by men

'Culture does not make people. People make culture. If it is true that the full humanity of women is not our culture, then we can and must make it our culture.'

– Chimamanda Ngozi Adichie, writer

WHEN I WAS editing *The Times* on Sundays, I used to have a running argument with the Night Editor, who put together the first edition. If a woman was the subject of a news story, he would always shout across to the picture desk, 'Is she photogenic?' I would protest, 'You never ask that of a man.' He shrugged. To him it was a given that women would appear in photos because they were eye candy, but men would appear because they were newsworthy. Think of the women you see on the front pages of even serious newspapers today. Chances are they will be either an actress or the Duchess of Cambridge.

Why *do* we assume that men deserve to have greater authority than women? Of course, centuries of patriarchy have instilled this belief in us, but it's also what the world around us, even now, is telling us every day. In this case, it was telling us that women are judged by their looks and men by their achievements. Media organizations are still mainly run by men, so when we turn on our televisions, read our newspapers or go online, the messages we receive are more often those that men deem important. The senior presenters, journalists and opinion writers who give us the news and interpret it are still mainly men. The experts they interview and quote are more likely to be men. And women who work in these organizations are still often disdained by their male colleagues.

We get the same message from movies and TV drama. Although this has started to improve in the past few years, with stronger female characters on screen, women still have less speaking time, are much more likely to be portrayed as sex objects or murder victims, and are overwhelmingly likely to appear in dramas directed by men.

Let's start with the news media. Turn on the TV to watch a news or current affairs show and, until very recently, almost all the people with gravitas, experience and authority explaining events to us were men. Women aren't allowed to appear on TV talking about news and current affairs once they have wrinkles. They're only permitted to present *Cash in the Attic* or *Antiques Roadshow*. Men, by contrast, can have faces that look like a relief map of the Mendips and still keep their jobs. As age and authority are connected in our minds, keeping older women off the screen helps to confirm our unconscious bias that equates 'male' with 'authority'. As we heard earlier, young women find it particularly hard to be taken seriously.

This has improved in the past few years. Laura Kuenssberg, the BBC's Political Editor, and Katya Adler, its Europe Editor, now have senior posts conveying authority. And Fiona Bruce has at last broken the all-male monopoly of chairing BBC1's *Question Time*. But they are still much younger – and therefore less obviously experienced and authoritative – than the distinguished men who surround them.

All over the world, we still see the tired old trope of the attractive young female presenter paired with the authoritative, avuncular older man. He reports the serious news and she is given the fluffier, human-interest stuff. I once pitched to the BBC the idea of a programme which I – then in my early fifties – would present alongside Amol Rajan, a man then in his early thirties (now media editor at the BBC). We had worked together well at the *Independent*. We thought it would be an amusing reversal of the old stereotype and, as well as being taboo-busting, it would be thought-provoking too. Although we got a polite reception from BBC executives, it was never commissioned.

Only 4 per cent of female TV reporters are over fifty, compared with 33 per cent of men.[1] Miriam O'Reilly, the BBC *Countryfile* presenter who was taken off screen at the sight of the first grey hair, won her age-discrimination case. It is, of course, sex discrimination too, as older men aren't sacked for the crime of being over fifty. O'Reilly was fifty-two when she lost her job, but her sixty-eight-year-old co-presenter, John Craven, remained on air.

This matters, as older women are full of insight, wisdom, experience – and authority. Post-menopause, they also tend to be bolder and braver. Elaine Chao, 67, told me, 'I am a lot more confrontational now. I think it's age. I just went into a meeting recently, in the Situation Room, I was the only one against this whole room. And someone said, I basically took on a bazooka and killed them all. I think if I were younger, I would be more reluctant. But now it's like, I don't care.'[2]

In most cultures, the wise older woman, the matriarch of the family, is respected and revered. If broadcasters assume that viewers cannot bear to see an older woman on screen, while we are apparently perfectly happy to look at an older man, they are depriving us of female wisdom. And they are entrenching the stereotype of the expert, authoritative man. This widens the authority gap.

It's not just presenters who give us the impression that men are more authoritative than women. It's who the media call on as experts. A study called 'Women, Men and News' found that only 16 per cent of stories about politics and government used women as sources.[3] The Global Institute for Women's Leadership analysed fifteen top news sources in the UK, Australia and the US, and discovered that for every mention of a prominent female STEM expert in a news story about coronavirus, there were nineteen mentions of male ones.[4]

At least part of this is lazy journalism. Ed Yong, science writer at the *Atlantic*, looked back at his articles over the course of a year and found that just 24 per cent of his quoted sources were women and that more than a third of his stories quoted no women at all.[5] 'That surprised me,' he wrote. 'I knew it wasn't going to be 50 per cent, but I didn't think it would be that low, either. I knew that I care about equality, so I deluded myself into thinking that I wasn't part of the problem. I assumed that my passive concern would be enough. Passive concern never is.'

He set about redressing this gap. Initially, he tracked how many people he asked for a quote, in case women were simply more reticent than men. They weren't. Then he started actively looking for female scientists who were just as qualified to comment on his stories as male ones. It wasn't that hard: he estimates it took him perhaps fifteen minutes more on each assignment to look beyond the usual suspects.

Crucially, he tracked what he was doing on a simple spreadsheet. 'I can't overstate the importance of that: it is a vaccine against self-delusion. It prevents me from wrongly believing that all is well. I've been doing this for two years now. Four months after I started, the proportion of women who have a voice in my stories hit 50 per cent, and has stayed roughly there ever since.'

And it's not as if the female experts were less qualified than the tried-and-tested male ones, he says. 'We don't contact the usual suspects because we've made some objective assessment of their worth but because they were the *easiest people to contact*. We knew their names. They topped a Google search. Other journalists had contacted them. They had reputations, but they accrued those reputations in a world where women are systematically disadvantaged compared to men.' So, by being lazy in relying on sources that have been amplified by a sexist media, we are just replicating and enlarging that bias.

It's not just in STEM stories either. When Women in Journalism tracked a week's worth of journalism in 2020, it found that only 16 per cent of people quoted on the front pages of newspapers were women.[6] Just one of those women was black.

At last (some) media organizations are recognizing this as a problem. After the *Financial Times* found that only 21 per cent of people quoted in the paper were women, it developed a bot to analyse pronouns and first names to see whether a source was male or female.[7]

Section editors are now alerted if they are not featuring enough women in their stories. It's perhaps no coincidence that the editor of the *FT* is a woman.

The BBC has recently started doing fantastic work in trying to level up the percentage of women it invites on to its programmes. 50:50 The Equality Project was started by one (male) journalist, Ros Atkins, after listening in his car to a radio programme that didn't feature a single woman, and has now spread across the BBC and beyond, to more than sixty other organizations in twenty countries. It involves counting the numbers of men and women in all the content they produce, with the aim of reaching 50 per cent women on screen, on air and in lead roles across all genres, from drama to sport to news. They count only contributors whom they can control, so they don't, for instance, include the prime minister or the only eyewitness to an event. But they do count reporters, analysts, experts, case studies and anyone else who is invited to appear. Crucially, they are determined to achieve equality without any drop in standards. The women they invite on air must be just as well qualified as the men.

The 2020 figures showed that, within the BBC, two thirds of teams had managed to achieve a 50:50 ratio, and of the teams that had been part of the project for at least two years, 93 per cent had more than 45 per cent female contributors. As Ed Yong found, tracking with data is the key, otherwise it's too easy to slip back into default male mode. As the authors of a study on the 50:50 project wrote: 'The one universal insight echoed by everyone we interviewed was the value of collecting their own data and following it over time. We heard this from people who initially argued that they were already doing a good job of representing women on screen, only to find themselves at around 30 per cent female representation after actually counting; from those who hit 50 per cent women quickly, and then followed their data and found themselves backsliding; and from those who had yet to hit the mark. Data serves as an essential reality check on one's gut instincts, countering overconfidence, sustaining motivation and encouraging goal pursuit.'[8]

These results show that an equal gender balance *can* be achieved, without undermining standards and without much extra effort. And it's not just a token exercise: the subliminal effect that it has on viewers and listeners will surely help to narrow the authority gap. For it is what we see in the outside world that shapes the unconscious bias in our brains. And, as we saw in Chapter 9, that bias can be reduced over time.

This move is very much needed. The Global Media Monitoring Project (GMMP) 2015 survey (it is conducted only every five years and the 2020 study was postponed because of the pandemic) found that, across the world, women comprised only 19 per cent of experts who were quoted in news stories, barely higher than ten years before.[9]

Even in areas which affect women far more than men, they are much less likely to be used as sources. A study of news coverage in the 2012 US presidential election found that on abortion, 81 per cent of the people quoted were men and, on birth control, it was 75 per cent.[10]

Women are also nearly four times as likely to be described by their family status as men are.[11] This rang a bell for me. Many years ago, I wrote a feature for *The Times* about a Kafkaesque experience I had with my local council: they took me to court for putting my rubbish out on a Tuesday when the collection day was . . . Tuesday. The *Sun* shamelessly stole the gist of the article and wrote it up as a news story. Had I been a man, the reporter would doubtless have described me as a journalist, or perhaps even assistant editor of *The Times*. Instead, I appeared as just 'Mary Ann, mum of two'.

It is journalists who decide what we see, hear and read, and they are still mainly male. In the UK and Ireland, 39 per cent of media professionals are women, only slightly higher than the global average of 37 per cent. As male journalists are less likely to quote female sources, this matters if we want to give women more visibility and authority.

And the disproportion is even greater at the top. Less than a quarter of the world's editors are female.[12] If men are running the media, the world reflected to readers and viewers will be seen through their eyes. As Eleanor Mills, Chair of Women in Journalism, puts it, 'Society sees itself not as it is, but through the prism of a predominantly old, white, privileged male gaze.'[13] She uses the example of the resignation of the Defence Secretary, Michael Fallon, after #MeToo finally led to widespread allegations of sexual misconduct by British politicians across Westminster.

'When political hacks meet, it's common knowledge who in Parliament has wandering hands. So why were these open secrets not reported before? Why were they not considered "news"? Why until Weinstein's outing changed the debate were other powerful men allowed to go on getting away with it? Could it be because some of the male posse that runs the media have similarly out-of-date views about what powerful chaps should be allowed to do?' In his resignation letter Fallon denied many of the allegations against him, but admitted that his behaviour towards women had 'fallen short'.

At that time, all but one of the political editors on British national newspapers was a man. Part of the reason that men outnumber women at the top of the media is that female journalists, like other professional women, are often sidelined once they have children. As soon as I fell pregnant, my boss suggested to me that I leave my job as opinion editor and become a jobbing leader writer. Fiona Pearson, a senior sub-editor at *The Times*, had a similar experience. 'On various occasions, I was asked by various male colleagues (bosses and managing editors) whether I had thought that it might be better if I chose to focus on my "lovely husband and wonderful children". There was also always an assertion that, as a mother, I should not put the hours in and my boss clearly felt threatened that I wanted to be in the office just as he was, rather than running home to make the kids their tea. I do remember asking one of these male colleagues if he had contemplated focusing on being at home with his lovely wife and wonderful children. He looked perplexed.'[14] If women are encouraged to step sideways once they become mothers, but men aren't once they become fathers, we are always going to have many more men at the top of organizations, and the authority gap will yawn as wide as ever.

But the sexism starts much earlier in female journalists' careers, even in these days of lip-service equality. Vicky, 28, a producer and reporter who has worked at a number of radio stations, summed up her media life for me. 'Here are some of the things I have experienced: being told I would never get to where I want to be, but I would achieve something as I am "lovely", being told that I am good at "light-hearted content", nothing else. I have been given light-hearted content to report on, because of my gender. I have had male colleagues smirk when I have asked for help, I have had them unload their work on to me and then get all the credit – and the pay for it too. I have been told that I have got the job because of my looks, I have been told I am an asset to the team because of my looks. I have had people assume I will always come out with something silly, being ridiculed for honest comments and questions, not getting career development and support as I work too hard and I am too enthusiastic. I have been mansplained how to do my job, I have been pushed out of opportunities for being a female. My work has not been credited or appreciated because I am a woman, but males' has been. I have been told that, "I hope there were no sexual favours involved" when a producer told a manager I had done him a favour.'[15] Quite a litany. And her experience is not unusual: female journalists do tend to get shunted towards what journalists call the 'fluffier' beats of health, education, features, celebrities, lifestyle and the arts.

And if women resist this pressure, they can find their job much harder than it is for their male colleagues. I spent three decades covering British politics and, in my earlier years, many Conservative

Cabinet ministers were members of the all-male Garrick Club, as were most of the political editors and columnists with whom I hung out. My friends and rivals would come back from the Garrick with great stories and gossip from their sources but, unfortunately, I had the wrong anatomy, which made joining their club impossible.

There was also a tradition of male political reporters and politicians bonding over playing golf and playing or watching football, which again I couldn't compete with. As more women came into politics (they made up only 3.5 per cent of MPs when I first started covering Parliament), they did make an effort to form a special bond with the female reporters and columnists. It was a help, but they were rarely in positions as senior as the men. And the only female after Margaret Thatcher who made it to prime minister – Theresa May – was famously the least gregarious politician any of us had ever met.

In France, female and minority ethnic journalists were not just excluded by default, as we were, but actively harassed by a clique of their male colleagues, who called themselves the Ligue du LOL. This group of about thirty members, some of them in very senior positions, spread pornographic memes of their female colleagues online and doctored photos to humiliate their victims.[16]

When female journalists do venture into traditionally 'male' territory, such as politics, business and sport, they are often treated with less respect. The authority gap there is wider still. Julia is a former digital journalist at a left-leaning magazine. She gave me an example: 'We had one female political reporter on a team of four. Her stories were often the most read. On at least two occasions, the editor of the magazine called me up and told me to change the splash [the lead story] from one of her stories because it was "not serious". He never made the request about anyone else's stories. Both she and I believed it was because she was a woman.

'On another occasion, the same experienced female reporter spent all day live-blogging the local council elections. The magazine editor came in and, ignoring her, asked the male intern to give him the lowdown on the latest local politics developments. When I asked him why he didn't speak to the person who was actually covering them, he belittled me.'[17] This sort of treatment is so damaging to women's self-confidence.

Being a black female political journalist is even harder – and rarer. Anne Alexander was the first black woman to join the lobby: the elite group of political correspondents who get to attend the daily Westminster briefing. Eighteen years later, at the time of writing, she is still the only black woman there. When she attended a parliamentary reception, she was talking to a group of people when an MP standing nearby 'suddenly turned in my direction, gave me a quick glance, then handed me his empty glass before turning away and continuing his conversation'.[18] He assumed she was a waitress.

We need more women like Anne Alexander in political journalism if we are to do anything about the middle-aged, middle-class, white male lens through which the media sees the world. As she says, 'My perspective on some issues and stories as a black woman from a working-class background will be different to that of a white man from a middle-class background.' When journalists are drawn from such a narrow pool, they are never going to understand the full breadth and depth of the country. No wonder so many were caught napping by the UK voting for Brexit or the safety breaches at Grenfell Tower.

Alison Kervin is sports editor of the *Mail on Sunday*, the first woman to edit the sports section of a national newspaper. She is constantly met with incredulity, especially from men, who simply can't believe that she knows what she is talking about. When she first started her job, she was having a coffee in the canteen with a man from the systems department, who was explaining to her how the expenses system worked. 'Someone nearby had a question about sport and was told, "Well, the sports editor's over there." So he came over and straight away asked the systems guy if he knew who was playing on Saturday. I said to him, "No, I'm the sports editor." He said, "Oh, right," and turned back to the guy again and repeated the question.' It just did not compute. 'In sport,' she told me, 'there's such an assumption about knowledge based on gender. There's this popular mythology about women not understanding the offside rule. In every job I've had, someone has said, "Do you know the offside rule? Go on, do it with these cups." '[19]

In another incident, she describes being in a big rugby pub where there were some men at the bar arguing about when the famous rugby player Jason Leonard won his first cap. Kervin had co-authored Leonard's autobiography. She takes up the story. 'One said, "It was definitely 1988," and the other claimed, "It was definitely 1989," and they were getting quite vicious with each other. I knew it was 1990, so I interjected, "Excuse me, I didn't mean to overhear, but it was 1990." They glared at me and said, "What would you know? No, it's not!" and carried on arguing with each other. I stood there at the bar, waiting for my drink, and then said, "It *is* 1990" as I walked away.

'One of them said, "No, it isn't. How the hell would you know? I *know* it's 1989 because I've just read his book." I had the glorious moment of turning round and saying, "I know it's 1990 because I *wrote* his book!" I was only trying to be helpful, but I don't think a man saying that would have just been dismissed.' When she wrote another book about the history of the Rugby World Cup, her publishers advised her not to use her first name, but to put 'A. Kervin' on the cover. She refused, and acknowledges that it probably sold less well as a result.

Kervin's favourite story is of the time she had a man round to do her plumbing. He asked her what she did. 'I told him I was a sports editor for a national newspaper. He said, "Blimey! You probably know as much about sport as I do!" '

Jess Brammar used to be acting editor of BBC2's *Newsnight* and is now editor-in-chief of HuffPost UK, holding down one of the

biggest jobs in digital journalism. She is fed up with the male dom-
ination of the media and the exclusionary feelings it engenders in
women. 'I tend to avoid journalism events and parties because I find
them quite excruciating. That is because I've so often been the person standing around with groups of men, being ignored. I'm just
sick of being ignored as a woman. I don't find it that enjoyable.'[20]

At *Newsnight*, women were routinely talked over at meetings.
Because she was senior, 'I would make an effort to see when women
were being spoken over in an editorial conference, and I would say,
"Hang on a minute. So-and-so was trying to speak."' She had to use
the same amplification strategy as the female aides in the Obama
administration.

Media organizations are notoriously competitive and there is a
lot of bullshitting and macho posturing. But I know, from decades
of sitting at morning and afternoon conferences at newspapers,
where the content is decided, how important it is to have diversity
at the top. If men run newspapers, they will tend to report the news
and choose features in a way that reflects their interests and priorities. They are more likely to dismiss as frivolous things that concern
women. I can remember male colleagues literally rolling their eyes
when I suggested stories about childcare or work–life balance.

All these things matter, not just for the female journalists involved, but because the media reflect the world back to us. If the
mirror is distorted in a male direction, we will continue to see the
world through a male gaze. And that will shape our reflex attitudes
and unconscious bias, which will perpetuate the authority gap.

What is most important, though, if we want women's views to
be accorded the same authority as men's, is for them to be equally
present in the opinion pages of newspapers, which are still, even in
these digital days, the crucible of public debate. When I began editing the opinion page of *The Times* in 1988, I inherited three regular
columnists a day, six days a week. Every single one of those eight-

een writers was male. I had to fight to be allowed my first woman on that sacrosanct page.

When I confronted my editor about it, he replied, 'Find me a woman who can write as well as William Rees-Mogg [the distinguished former editor of *The Times*, who still had a weekly column] and you can give her a slot.' The argument, he thought, was over.

'OK,' I retorted. 'How about Libby Purves?' She was already writing excellent serious opinion pieces that were tucked away on the 'fluffy', 'female' features pages. To his credit, he agreed, and she is still writing a weekly column there some thirty years later, the only one of that generation to have lasted. Yet even in 2020, five out of the six lead columnists at *The Times* are men, and on half the days of the week, there are three regular male columnists to one female one. What does that tell us about how seriously we take women's opinions compared to men's?

It's not difficult to find women who write brilliantly. The *Guardian* (which is edited by a woman) has as many female op-ed columnists as male ones, including award-winning writers such as Marina Hyde and Nesrine Malik, and serious political pundits such as Polly Toynbee. But, like Ed Yong, with his scientific sources, editors may have to go out and look for them. Because of all the social pressures they experience, women are less likely than men to push themselves forward.

This is true in all organizations. To achieve equality, we have to take conscious measures, and keep taking them. We have to keep checking that we are giving women an equal chance and that men aren't elbowing them aside. We have to remember that women aren't allowed to elbow men back without suffering the consequences. If we don't keep at it, society snaps back to the male default, like an elastic band as soon as you take the pressure off. And this has nothing to do with men being better than women, but about the biases that shape our behaviour and our attitudes to the two genders.

When women with authority *are* written about in newspapers, it's often in a way that undermines their authority. Deborah Cameron and Sylvia Shaw analysed the press coverage of female politicians and TV anchors during the 2015 general election campaign in their book *Gender, Power and Political Speech.*[21]

As Cameron writes, 'The women being described here had featured prominently in a debate watched by millions; one of them also had a day job running a small country. And what did the pundits compare them to? Head girls, primary school teachers, headmistresses, nurses, matron. This is how female authority is made intelligible: through allusions to a set of archetypal roles in which women have traditionally exercised power – prototypically over children, or over adults infantilised by illness. There was no pattern of analogous references to men: their authority in the political sphere is taken for granted, and does not call for comment or explanation.' And, of course, the subtext is that any man who accepts female authority has thereby been infantilized, has reduced himself to the status of a child.[22]

The disparaging references remind us of *Carry On* films, *St Trinian's* and P. G. Wodehouse's notorious aunts. Cameron explains: 'In such instances women's authority is both a joke and a threat, or perhaps I should say, it's made into a joke to defuse the threat: they're bossy boots, petty tyrants, and in popular culture often grotesque – ageing, physically unattractive and either sexless or pathologically oversexed "man-eaters".'

So if newspapers still portray an outdated image of women, and marginalize them both in their newsrooms and on their pages, what about advertising? Ads, after all, are also supposed to reflect our world back to us, to give us an idea of how normal people and families act and relate to each other. They are another lens through which we frame our lives. If anything, the story here is even worse. A Channel 4 survey looked at the 1,000 most-watched TV ads and found that 41 per cent of them showed women as housewives and only 28 per cent as office workers.[23]

This isn't just a British problem. A 2018 poll of men and women from twenty-eight countries found that 64 per cent thought advertisers should do more to eliminate traditional or old-fashioned roles of men and women in their ads.[24] Nearly half said they still saw sexist ads that offended them. And three quarters said that they feel more positive towards companies whose ads portray men and women with the same abilities and roles. So consumers really want change, and the advertisers are changing too – but they haven't quite kept up.

In Britain, the Advertising Standards Authority has finally banned the use of harmful gender stereotypes in advertising. This is terrific, but the scope is smaller than you might imagine; the emphasis is on the word 'harmful'. Advertisers are no longer allowed to portray a woman looking helplessly at a car engine, or indeed a man looking helplessly at a dirty nappy. But that doesn't mean women can't be portrayed as housewives, unnaturally excited by the power of a kitchen cleaning spray, or men as executives, powering into the office, briefcase in hand, while the young PAs swivel their necks to watch the boss stride past.

Perhaps the most insidious effect on our internal stereotypes of women's and men's roles comes from the stories we are told by dramas on film and TV. Thankfully, here there has been huge progress of late, particularly on TV. Dramas such as *Fleabag*, *Killing Eve*, *Gentleman Jack*, *I May Destroy You* and *Big Little Lies*, written by women, directed by women, with really interesting, complex characters for women of all ages, are as refreshing as a cold beer in a heatwave. And they have been successful too, which makes one wonder why TV executives took so long to commission them. When the second series of *Killing Eve* was put up on the BBC's iPlayer, 2.6 million people downloaded the box set in thirty-six hours.

By portraying women as powerful, complicated, morally ambiguous, funny and commanding, they challenge the stereotypes
that older people have grown up with, and provide role models for
a whole new generation of girls and young women.

This is such a new phenomenon. Over the decade to 2016, just
11 per cent of all UK films and 28 per cent of all TV episodes were
predominantly written by women.[25] And female writers were pigeon-
holed into soaps and children's programming, finding it hard to
move into prime-time drama, comedy or light entertainment.

So often the parts were two-dimensional too. Laura Bates was an
actor before she founded the Everyday Sexism Project in 2014. Before going to auditions, she would be sent a 'casting breakdown',
which was supposed to give context to the character for whom she
was trying out. 'My boyfriend was an actor at the time as well,' she
told me, 'and he would get casting breakdowns that were long and
detailed and told you about the character that he was going up for
and how they were shy and introverted and how they had had a bad
childhood experience that made them this way and so on. I once got
a casting breakdown that was four letters long: it said, "32DD", that
was it. Or I'd get these breakdowns that said, "She's sexy, but virgin-
al," or "She's naive, but fuckable," just these incredible stereotypes.'[26]

Moira Buffini, playwright, director and screenwriter, agrees that
television is changing, although there is more work to do. 'It has
been a world populated by mature and interesting men, and young,
beautiful women, and that just has to stop. It has an insidious effect
on people. Female characters need to lead the action, as they do in
their lives, where they are not mere helpmeets.'[27]

She is right about how powerful the influence of TV role models
is. The Geena Davis Institute on Gender in Media decided to find
out whether the character of Dr Dana Scully, played by Gillian
Anderson in *The X-Files*, changed girls' and women's attitudes to
STEM subjects.[28] Scully, a medical doctor turned paranormal de-

tective, was one of the first really interesting female characters in a STEM field to appear on popular TV and the first to play a leading role. She was known for her brilliance, her strength, her objectivity and her scepticism.

Their findings were quite extraordinary. Taking a sample of more than 2,000 women over the age of twenty-five – all old enough to have watched *The X-Files* – they found that 63 per cent of the women who worked in STEM cited Scully as their role model. Of the women who were familiar with Scully, half said that she increased their interest in STEM and 63 per cent said that she increased their confidence that they could excel in a male-dominated profession. Regular watchers of *The X-Files* were 50 per cent more likely than light watchers to have worked in a STEM field.

If just one strong female character in one popular television show can have that large an effect on a generation of girls and women, think how much difference it would make to have a whole lot more. These changes aren't just token or cosmetic; they are truly life-changing.

The world of film is also improving for women, but more slowly than television. In the hundred top-grossing films of 2019, 43 per cent had a female lead or co-lead, up from 20 per cent in 2007 and 32 per cent in 2017.[29] But 42 per cent of all films in 2019 still failed the Bechdel Test, which only asks for two named female characters to have a conversation that isn't about men.[30] That's not a high bar. And, on average, there were still two men on screen for every one woman.[31]

In the hundred top-grossing films, there were only three female leads or co-leads aged over forty-five and only eleven women of colour. And just 16 per cent of them, in 2020, were directed by women – though that is a big increase from 4 per cent in 2018.[32]

Korean-American director Jennifer Yuh Nelson is one of just a handful of women who have been allowed to direct an animated film, where the disproportion is greater still. She's directed *Kung Fu Panda 2* and *3* and *The Darkest Minds*. But still, she told me, she bumps up against the old stereotypes. Only thirteen women of colour have directed any of the 1,300 top-grossing movies between 2007 and 2019. Yuh Nelson is one of only two (the other being Ava DuVernay) who has directed more than one.[33] She explains, 'I'm kind of one of those walking anomalies where people are saying, "Wow, you're a woman doing this, that's rare, you're an Asian doing this, that's rare." It's so many levels of rare and it's all stuff I can't help. I can't suddenly stop being Asian, I can't suddenly stop being a woman.'

So she often finds that people are surprised when they meet her. For one job interview, she was asked to meet her potential producer in a hotel lobby. He walked past her several times, even though the lobby was otherwise empty and she was holding her giant animation portfolio. He was sure the storyboard artist he was supposed to meet must be hiding somewhere.

And now that she is a director, the same thing happens. 'I'll be sitting in the room by myself and the actor will walk in. They're going to see the director and they literally look at me, look confused for a second and you can tell they're calculating in their head, "Is this somebody that I'm supposed to be meeting?" And they do a double-take. Or if I'm standing next to a man, they'll do a double-take and look at the man first and then look at me and look very confused. People, whether they mean to or not, just have to take a moment to realize that I am the director.'

So why are there still so few female directors? The quality of the movies they make is just as good.[34] It doesn't even make commercial sense. Films made by women are much more likely to have important female characters. And films with female leads do better at the box office. For the highest-grossing films between 2014 and 2017, female-led ones outperformed male-led ones at every budget level.[35] Movies that passed the Bechdel test also outperformed those that didn't. In fact, all the films that grossed more than $1 billion globally in that period passed it.

Since #MeToo, there has definitely been a push to have more strong female characters on screen and more senior women behind the camera. The result has been great movies such as *Three Billboards outside Ebbing, Missouri*; *Hidden Figures*; and *Late Night*. Studios are at last being shamed into giving women a better chance. This can only be good news for those of us who want to see women depicted on screen as nuanced, rounded characters showing agency and authority, not just as sex objects, supportive wives or murder victims.

But behind the scenes, women still face resistance from men. Kitty, a producer in her early thirties, eventually left her job in Hollywood because her male boss, outwardly liberal and progressive, said things like, 'I find it incredibly irritating how much you care about female characters.'[36] He refused to take on an excellent film project that she put forward. 'It's about consent,' she told me, 'and is a consciously feminist film. That was a clincher for me. He saw how well-written the script was, but it just rubbed him up the wrong way.' The script he passed on, *Promising Young Woman*, by Emerald Fennell, of *Killing Eve* fame, was taken up by Margot Robbie as producer, starring Carey Mulligan. It was nominated for five Academy Awards.

Kitty is excited by the progress now being made, but still thinks that life is harder for women than for men in Hollywood. 'In order to be a successful female film-maker you have to be the absolute best,' she says. 'In order to be a successful male film-maker, you can be mediocre and you'll be hired again. The male producers and film-makers and distributors are still running Hollywood and they want to put women on lists. But then the women's names get filtered out. It's as if there's an invisible hand, cupped, waiting for females to fail.' This is true in so many fields. As Helle Thorning-Schmidt said, we'll know we've achieved true equality when there are as many mediocre women as mediocre men in authority.

★

I've talked about how the news media, TV and films shape our view of ourselves and our place in society, but what about the power of religion? For those people who have faith, religious teaching is the framework by which they are supposed to lead their lives. And if their religion is telling them that men are superior to women, they are not just going to display unconscious bias, but probably the conscious variety too.

All the world's major religions teach some version of the golden rule. In Christianity, it is 'Do unto others as you would have them do unto you.' Buddhism advises, 'Hurt not others in ways that you yourself would find hurtful.' Hindus have, 'One should never do that to another which one regards as injurious to one's own self.' And Mohammed instructed, 'As you would have people do to you, do to them; and what you dislike to be done to you, don't do to them.'

Yet most religions oppress women, discriminate against them, or put them in positions of inferiority that men would hate if it were done to them. How has this happened? The only explanation for this hypocrisy is that the cultures prevailing at the time these religions were founded were so patriarchal that oppression of women seemed natural – just as many Christians at one point saw no problem with slavery.

The trouble is that the world has moved on, but the scriptures haven't. So, for instance, the Koran still gives daughters half the inheritance of sons and says that a woman's testimony in court – at least in financial matters – is worth only half that of a man's. Similar verses in the Bible say that women shouldn't hold positions of leadership in the Church.

Women and girls have been discriminated against for too long in a twisted interpretation of the word of God. Not my words but those of former US President Jimmy Carter.[37] We all knew him as a devout Southern Baptist, but he reluctantly left his Church after sixty years when it said that women had to be subservient to their husbands and that they couldn't become deacons, pastors or chaplains.

Carter points out that women often served as leaders in the early Christian Church, but that, after the fourth century, men, he says, 'twisted and distorted Holy Scriptures to perpetuate their ascendant positions within the religious hierarchy . . . The truth is that male religious leaders have had – and still have – an option to interpret holy teachings either to exalt or subjugate women. They have, for their own selfish ends, overwhelmingly chosen the latter.'

This has allowed men to justify their control over women, in the name of a higher authority, for millennia. It has allowed them to restrict girls' and women's opportunities, to force them to marry older men, to prevent them competing for authority and power with men, and even to blame them for men's sexual weaknesses.

Mary McAleese is so incandescent about the refusal to grant women authority in her own Catholic Church that after she left office she got a doctorate in canon law. Why? I asked her. 'The exclusion of women in my Church greatly bothers me. Because it represents 1.2 billion people, one in six people in the world, I feel I can't walk away. My own Church is an empire of misogyny. It has carried down the 2000-year conduit attitudes that are still embedded in societies all over the world.'[38]

While she says she largely encountered respect in the world of politics, 'From my Church, especially in the higher echelons, I found resentment that I expected to be taken seriously, resentment that I had achieved a platform from where I could speak and be listened to.

'The same people who were arguing for the human rights and civil rights of Catholics [in Northern Ireland] didn't notice that within their own Church, the human rights and civil rights of women were routinely overlooked and neglected, quite deliberately of course.'

In the Church of England, too, there has been a huge struggle to accord women the same authority as men. The Movement for the

Ordination of Women, to allow women merely to become priests, battled tenaciously for nineteen years until finally, in 1992, the motion was passed by just two votes among lay people in the General Synod. This reform proved so contentious that 430 priests resigned and claimed compensation. It took a further twenty-two years for the Synod – the Church's governing body – to vote in favour of allowing women to become bishops.

Dame Sarah Mullally moved across to the Church from another position of authority: she was the government's chief nursing officer. The Diocese of London has famously been a hotbed of opposition to women priests: 13 per cent of her clergy don't accept women as priests or bishops and have to have special provision from 'flying bishops'. So it was a surprise when she was appointed to be Bishop of London, she told me: 'a surprise by the Church, but a delight by the world'.[39]

'Since I've been a bishop,' she says, 'I've been very much more conscious of my gender. I stand at the back of the church and as people leave they will say things like, "Oh, I didn't know what to expect from a woman, but you were all right," or, "You're actually quite pretty really," and I know they're trying to be nice but actually it's a very odd thing because I feel I'm doing it because I'm doing it, and not because I happen to be a woman.'

Many churchgoers still find it difficult to associate 'authority' with 'woman'. When Revd Lucy Winkett, the first female priest at St Paul's Cathedral, was celebrating Mass, a bewildered visitor asked a verger, 'Why is that priest speaking with a woman's voice?'[40] Mullally remembers visiting churches in Devon in her early years as a bishop. 'I'd turn up to a church service and they wonderfully save a parking space for the bishop, so you turn up, remove the cones, park the car and then suddenly a man comes running out and says, "I'm afraid this space is reserved for the Bishop," to which I say, "Well, I *am* the Bishop," to which he says, "Well, you're a woman." "I know I am," I reply, "but I'm still the Bishop."'

Some of this is an understandable difficulty in getting to grips with the new. But there is also blatant hostility from others to the idea of a woman wielding authority in the Church. Mullally says, 'I receive vile comments at times, particularly on social media. I tend to get more unpleasant things from people who are anonymous, rather than those people that I directly speak to. The challenge for me is to work out what is somebody's theological objection to me and what is discrimination.

'And it's not always easy, but sometimes it's absolutely, blatantly clear. So I do front some of that up, which I think people are often surprised at, but I think if you don't confront it, it's harder. You won't change it. People have been surprised to have that conversation; it's been disarming for people because they think we'll avoid it or not talk about it.'

Judaism allows female rabbis in its liberal wing, but not in the Orthodox wing, at least in the UK – though some have been allowed in the US. And ultra-Orthodox women have to wear wigs and bear as many children as they can. Gender segregation in the ultra-Orthodox community is rife. But the Hebrew Bible at least contains examples of wise, authoritative women such as the Witch of Endor, Sarah, Rebecca, Rachel, Deborah and Leah.

These women, though, are afforded nothing like as much authority as the men in the Bible. As Baroness (Julia) Neuberger, formerly senior rabbi at the West London Synagogue, explained it to me: 'They are women and they play a role and they're not as subservient as you find women in later religious literature. But do they have authority? Well, yes, to some extent, but not huge.'[41]

In traditional Jewish law, a man can divorce a woman, but a woman cannot divorce a man. But women have always had the right to own property, and throughout Jewish history they were encouraged to read and write and to run businesses while their husbands studied, so in some ways they had more freedom than most Christian women.

In Islam, the bias is blatant. Women have to sit separately from men in mosques and, in most cases, aren't allowed to be imams. Mosques are run almost entirely by men and so are Sharia councils. The 'voice' of the Muslim community is almost always male, though the recent election of Zara Mohammed as Secretary General of the Muslim Council of Britain is a huge step forward. Teenage girls are occasionally forced into marriages with much older men. Women have to dress much more modestly than men. In many communities, men – whether they are fathers, husbands or brothers – expect to take decisions about how the girls and women in their families lead their lives.

Gina Khan, a brave Muslim woman who lives in Birmingham, decided to speak out about this, and had a brick thrown through her window for her pains. 'We are just treated as second-best,' she told me when I went to visit her there. 'It has always been like that. It does not matter whether you are from a village and backward or from a cultured Asian family – the mentality across the board is the same.

'You are fighting this mentality all your life, so it is hard to be who you are. You can either be miserable, as I was for thirty-four years, or you can challenge it to say, "You know what? I am a human being, God gave me a brain equal to the brain he has given you and I am not going to bend over and pray behind you just because you are a man."

'Muslim women aren't supposed to make waves. I didn't even hear my own screams and tears for thirty-four years. I have now stepped back and decided to understand and challenge my religion.'

Sharmeen Obaid-Chinoy is an unflinching documentary-maker who won two Academy Awards for the films she has made about the treatment of women in Pakistan. Women's rights have regressed there, she told me, thanks to men's interpretation of Islam. 'The regression in women's rights is directly linked to the Islamicization of this country. So while the religion grants access and rights to women, the interpretation of the religion is articulated by the men in this country. They believe that the religion does not give women the rights that the women claim that it does. And so religion is being used and distorted by men so that they can suppress women. Men in this country take out their frustration on women. They believe that women are inferior, that women should spend their entire life serving them in their homes. It has nothing to do with religion, but it has been manipulated in a way that people believe this is their religious right.'[42]

Interestingly, an experiment in Egypt showed that when Muslim people were presented with a brief religious argument interpreting a Koranic verse in favour of women's leadership, they were 24 per cent more inclined to accept arguments for women in senior political leadership positions, such as Prime Minister or President.[43] And this was more effective than using a non-religious argument based on scientific studies. So maybe a more liberal interpretation of the Koran can be used to achieve change in Muslim communities. But that requires male Muslim leaders and imams to believe in greater equality between the sexes and then to preach the value of it.

Jimmy Carter was right. It is selfishness – the determination of men to hold on to a better position at the expense of women – that has allowed continued discrimination against women in almost all religions. Which is curious, because the sin that the golden rule was designed to thwart is selfishness: don't put yourself first, but do as you would be done by. All religions preach that selfishness is wrong, yet their institutions protect some men's selfish wish to cling to power and exclude women from positions of authority. And while that is the case, believers will have the old stereotypes drilled into the core of their being.

Men will feel justified in believing – which suits their interests – that women are inferior and subordinate. Women, meanwhile, will be torn between the religious teachings they imbibe and a feeling that the world shouldn't have to be that way.

*

So, to sum up, we still live in a world that is framed by men, which casts men in a position of superiority and authority and women as subordinate beings. It is a world in which men, much more than women, are cited as experts. And it is a world in which men get almost all the best parts. These messages, which surround us every waking hour, are absorbed by our brains and turned into stereotypes. We then apply these stereotypes to real women and men; and that is what perpetuates the authority gap.

There *is* progress in some areas. TV and films are starting to give us a broader, more nuanced view of what women can be like, with many more strong female characters in positions of agency and authority. The 50:50 project is dramatically improving the visibility of expert and authoritative women. Advertising is slowly getting better. Hollywood is hiring slightly more female directors. Actresses no longer have to be confined to being sex objects in order to get on screen, and older women are increasingly being cast in films.

There has recently been a rash of new female editors of national newspapers in Britain, from Emma Tucker at the *Sunday Times* to Victoria Newton at the *Sun*. If they can change the traditional – sometimes toxic – male culture on their papers, that should help to narrow the authority gap.

Religion, as always, is dragging along at the back. The Church of England has shown that it is possible to reinterpret the Word – but only after decades of struggle, and there are still many men (and a few women) in the Anglican Church who are uncomfortable with women in authority. Many people of faith believe that it is categorically not the role of religion to move with the times; that religion is there as a counterbalance to liberal modernity and that the Scriptures must prevail. While this remains the case, women will be stuck with oppressively patriarchal practices that were followed thousands of years ago. It is hard to see their liberation coming soon.

Lady Macbeth meets Medusa
Why do we hate women in power?

'It's not easy to be a woman in politics. That's an understatement. It can be excruciating, humiliating. The moment a woman steps forward and says, "I'm running for office," it begins: the analysis of her face, her body, her voice, her demeanor; the diminishment of her stature, her ideas, her accomplishments, her integrity. It can be unbelievably cruel.'

– Hillary Clinton

JADRANKA KOSOR, CROATIA's first – and so far only – female prime minister since it became a democracy, came to power in 2009 when her predecessor stood down suddenly. I went to visit her in Zagreb and found her still bitter about the whole experience. She believes that a vile thread of misogyny ran through her treatment by both politicians and the media. From the start, they were convinced that she would be useless. They couldn't believe that a woman could lead them out of their economic and diplomatic travails.

'The very moment I became Prime Minister, I was met with derision and disbelief,' she told me. 'How was *she* going to be able to do this? At the time, Croatia was in the throes of the financial crisis. GDP had fallen by almost 10 per cent and I was met with disbelief. People were saying, "She won't be able to get us out of this crisis, she'll lead us into complete failure." The day when I was chosen in the Parliament, there were many insults hurled at me, particularly regarding me as a woman, not as a politician.'[1]

The derision spread beyond Croatia to the European Union, where negotiations for Croatia to join the EU had become gridlocked. 'At the time, the head of the European Commission was José Manuel Barroso, and during our first meeting he also expressed disbelief. Slovenia and some other European countries were blocking negotiations for Croatia joining the European Union. Mr Barroso literally looked me up and down and asserted that my predecessor, who was a man, couldn't do it, so he didn't think there was any way I could.'

Within three months, Kosor had unblocked the negotiations and secured Croatia's entry to the EU. She turned out to be a tough and skilful negotiator, and formed an unexpectedly successful relationship with the Slovenian prime minister, who had previously fought to keep Croatia out of the EU.

If we are resistant to according women equal authority for their expertise, it's as nothing compared to our resistance to women exercising authority in the form of power. Although I am mainly focusing on female political leaders in this chapter, there is a similar pattern to our attitudes to women with power in any field.

You only have to look at Hillary Clinton's experience running against Donald Trump in the 2016 US election to see the extent of society's toxic misogyny when there was a prospect of having a female president. Hillary Clinton was publicly criticized for all manner of things: her voice was wrong, her hair was wrong, her clothes were wrong, her manner was wrong, she was unlikeable, she was cold, she was nasty, she was untrustworthy, she was crooked. Despite her being possibly the best-qualified presidential candidate ever on paper, any excuse was used to demonstrate that she was, to put it simply, just too *female*. She also had the added difficulty of galvanizing huge crowds at rallies without being accused of 'screaming' or 'screeching', dominating a debate without being accused of being 'angry', 'nasty' or 'aggressive', and looking commanding on camera while remaining warm and feminine. Disgusting tropes were used on Republican merchandise, including a mug depicting Trump holding Clinton's severed head as if it were Medusa's, a badge saying 'KFC Hillary Special: 2 fat thighs, 2 small breasts, left wing' and T-shirts with slogans such as 'Trump that bitch!' and 'Hillary sucks – but not like Monica'.

And the same pattern was played out all over again in the Democratic primary field in 2020. The adjectives that were used against Clinton – 'abrasive', 'shrill', 'aloof', 'cold', 'unlikeable' – were reprised in articles about Senators Kamala Harris, Elizabeth Warren, Kirsten Gillibrand and Amy Klobuchar.

These are classic examples of the status incongruity problem. As we learned in Chapter 6, the 'masculine' qualities required in a leader make us feel uncomfortable when they are displayed by a woman. As a result, we are more likely to dislike female than male leaders, and we penalize women for showing the very leadership traits that we reward in men. Men are dominant, women are abrasive. Men are powerful, women are power-seeking. Men are decisive, women are strident. As the American feminist Gloria Steinem once said: 'A man can be called assertive if he launches World War Three. A woman can be called assertive if she puts you on hold.'[2]

Caroline Heldman, a professor at Occidental College in Los Angeles, studies our attitudes to women in power. 'Part of our cultural DNA is this idea that men are supposed to be in charge,' she told me. 'So when women seek power, both men and women alike have an aversion to that. We tend to dislike or hate power-seeking women. The hatred is more from men than from women, but as it's a cultural norm, both men and women hold this aversion to power-seeking women.'[3] She finds that about a third of women (most of them white) and about two thirds of men have this bias against women who seek power. And strong Republican supporters, both male and female, were three times more likely than strong Democrats in 2018 to say that men are more emotionally suited to politics than women.[4]

This leads to another aspect of the authority gap: we are more inclined to accord authority in the form of power to men than to women. Men in particular are more reluctant to vote for female candidates. A lot of them don't like the idea of their country being run by a woman.

Ségolène Royal was all set to become France's first female president in 2007, but was narrowly defeated, she says, thanks to widespread misogyny. Men in politics believed they had what she calls 'property rights over the highest post in the state' and couldn't tolerate a woman winning the election.[5] Even men in her own party, the Socialists, undermined her. 'But who's going to look after the children?' asked Laurent Fabius, a former prime minister, when she announced her bid for the presidency. Other socialists suggested that she was lightweight, erratic and unintelligent. She was described as a 'nude dancer', a 'washing powder advertisement' and 'super-nanny', none of which would be wielded against men.

It was Julia Gillard, Australia's first female prime minister, who made the famous 'misogyny' speech in 2012 that echoed round the world. In it, she said, 'I was offended when the Leader of the Opposition [Tony Abbott] went outside in the front of Parliament and stood next to a sign that said, "Ditch the witch." I was offended when the Leader of the Opposition stood next to a sign that described me as a man's bitch. I was offended by those things. Misogyny, sexism, every day from this Leader of the Opposition.'

Her male colleagues were shocked by how she was treated when she was prime minister. Mike Rann was premier of South Australia and president of her Labour Party. On a visit to London, we met to talk about Gillard's experience, and I was struck by the extent of the solidarity that this middle-aged man felt for his female colleague. 'It was a disgrace,' he told me. 'I don't think I've seen anyone more denigrated, despite her incredible abilities, because of her gender. She was diminished because of her clothing, because of her looks, because she wasn't married and because she didn't have children. And so we saw an attempt by large segments of the media and by her political opponents to try to delegitimize her as prime minister because of her sex.

'There were constant references to the size of her backside, her clothing, her make-up, her hair; none of the things that we blokes had to put up with. I'm overweight. No one ever said that! Around the Cabinet room, the men were all overweight.

'What it showed to me is that female politicians can't win. If they don't have children, they're somehow inadequate. If they do, they should be at home looking after them. If they are attractive, they're not serious, they're dolly birds. If they're not attractive, they're also denigrated. So for women in politics, not just in Australia but all over the world, you need a much bigger dose of resilience to put up with this crap.'[6] And that's without taking into account the hideous trolling and threats of violence that female politicians receive on social media.

Gillard has become famous for calling out the sexism she experienced. But with hindsight, she told me, she waited too long to do so. 'In my initial phases of being prime minister, I decided I would ride it out because I thought that it was an early phase and the longer I was prime minister, the more the nation would get used to there being a female prime minister, and that these sexist critiques would diminish over time. What I actually found was that they grew over time. And because I'd left it and not called it out early, it was harder to call it out later. So the lesson from my own experiences is call it out early.'[7]

Australian politics is notoriously rough and tumble, but an academic analysis comparing the media treatment of Julia Gillard and another prime minister, Malcolm Turnbull, both of whom challenged rivals in their own party to become leader, shows the stark difference between the two.[8] When Gillard toppled Kevin Rudd to become prime minister, nearly 50 per cent of the articles contained words such as 'murderer', 'back-stabbing', 'knifing', 'decapitation', 'brutal', 'ruthless assassination' and 'execute'. She was compared to Lady Macbeth. When Turnbull toppled Tony Abbott, only 12 per cent of the articles were negative. He was 'brilliant', 'successful', 'clever', 'cunning', 'ambitious', had 'political skills' and 'gravitas' and had 'taken back the reins'.

When there were six women running for the Democratic presidential nomination in 2019–20, many of us hoped that, at last, the media would distinguish between them and not just use lazy stereotypes to describe them. As Hillary Clinton put it to me in 2019, 'Your book could not be more timely, because we have three women who are really in contention now for the potential of getting the nomination, and they're different. So it's not just one woman standing alone on the stage as I did in 2016. They look different, they dress different, they talk different.'[9] But she still feared that voters' hostility wouldn't change. 'I always get a little bit of a chuckle because in 2016 people would be questioned, "Would you ever vote for a woman?" and they'd say, "Of course I would – just not *that* woman," meaning me, and now, in 2019, we had enough women for a basketball team running and the same answer. People were asking the same question and they'd say, "Of course – just not *those* women."' And sure enough, it was the slightly doddery old white man who won the nomination.

Why? During the Democratic primary campaign, a *New York Times*/Siena poll found that 41 per cent of people who supported Joe Biden over Elizabeth Warren agreed with the statement that 'most of the women who run for President just aren't that likeable'.[10] What are the chances that not just Clinton, but all six of the women running for the nomination four years later were genuinely difficult to like? Could it tell us more about our attitudes to women seeking authority than about them as individuals?

As we saw in Chapter 7, if a woman is thought to be highly competent, she will also be judged to be less likeable than either similarly competent men or less competent women. As only highly competent women are likely to run for the presidency, for many voters they will be automatically unlikeable, particularly when you add ambition or 'power-seeking' to the mix. As Deborah Cameron puts it, 'The problem isn't that women who run for high office have particularly unattractive personalities, it's that we don't tend to like women who run for high office.'[11] And if we don't like them, we are less likely to vote for them, which means there will always be a gap between men and women when it comes to wielding authority – unless we can get over our prejudice.

The differential dislike of 'power-seeking' women is striking. A study on the backlash against female politicians found that 'participants experienced feelings of moral outrage' such as contempt, anger and disgust when women politicians were described as power-seeking.[12] In contrast, 'When participants saw male politicians as power-seeking, they also saw them as having greater agency (i.e., being more assertive, stronger, and tougher) and greater competence.'

Google 'Biden' and 'ambitious', and you will read about his 'ambitious immigration bill' or his 'ambitious climate plan'. Google 'Hillary Clinton' and 'ambitious' and you get 'pathologically ambitious', 'burning ambition' and 'the curse of Hillary Clinton's ambition'.

And it's not just men who don't like women who seek power, but women too, albeit in lower numbers. In 2016, a majority of white women (and substantially more white men) voted for Trump, a self-admitted sexual assaulter, even though there was a highly qualified woman on the ballot. How do we explain that? I asked Caroline Heldman. 'White women have voted Republican along with white men since the civil rights movement. It's partly to do with race. But it's also to do with the patriarchal bargain. This describes the idea that, to get along in a system with rigid gender rules, women will disadvantage their group identity or interest to get what they can out of the system individually. In the case of Republican women, the patriarchal bargain has a lot to do with relying on men for economic resources and male validation.

'For many women, especially white women with a lower level of education, it's threatening to see a woman pursuing power, as it's an implicit indictment of traditional gender roles when a woman seeks a position of power that a patriarchal society defines and assumes is male. We tend to think of patriarchy being held up by men, but in the end over 90 per cent of homemakers are women and they are the conduits passing down social norms about gender.' So if children grow up in a socially conservative household, they will imbibe the notion that men should be in charge over women and that mothers should stay at home. The authority gap will be baked into them from birth.

Two studies after the 2016 US election found that hostile sexism – agreeing with statements such as 'Most women interpret innocent remarks or acts as being sexist' and 'Many women are actually seeking special favours, such as hiring policies that favour them over men, under the guise of asking for equality' – was a very powerful predictor of voting for Trump, second only to partisanship.[13] [14] This wasn't the case in 2012, when the candidates were Obama and Romney, suggesting that fear of a woman president was a very important factor in 2016.

It is much harder for women in a presidential system. As prime minister, it isn't quite so important to be likeable as it is for a president. A prime minister is only head of government, not head of state. He or she doesn't have to exemplify the nation, to make it feel good about itself, to make it 'stand tall', in Ronald Reagan's words. In Britain, we have the Queen for that. In America, by contrast, the president is both head of government *and* head of state. And the president, unlike the prime minister, is directly elected. Voters often choose the kind of guy they would like to have a beer with – which is hard if you are not a guy and you don't much like beer.

There is an added wrinkle, as Hillary Clinton put it to me: 'In a parliamentary system, you run in your constituency, people get to know you and you have to maintain your position in that constituency and then you get to know the fellow members of your party and they can actually see you as a person, as a political actor, and they can make judgements about your leadership.

'In a presidential system, you have to start from the very beginning, men and women. But the skills that are required are still more difficult for women. Raising all that money is hard for anyone; raising it as a woman is a little bit more difficult. Getting coverage that is not sexist through implicit bias is also more difficult.'[15]

In Britain's parliamentary system, it *has* been a little easier. Prime ministers are not directly elected: they win their job because they are the leader of the Party that has the largest number of MPs in the House of Commons. The Conservative Party elected both Margaret Thatcher and Theresa May as its leaders, and Thatcher won three elections in a row.

Thatcher came up against a huge amount of sexist prejudice in the 1970s and 1980s, and fought it head on. Initially, people thought a woman would be too weak to lead. How wrong they were! She didn't try to soften her approach and use warm authority to head off the hostility to her: indeed, she relished her nickname, 'The Iron Lady'. (In private, though, she was quite capable of flirting with her favoured colleagues.)[16] She might have been widely disliked, but voters still admired her, and she was hugely successful electorally, helped by the opposition being in disarray. In 2008, she was voted the third-best post-war prime minister, after Churchill and Attlee, in a BBC *Newsnight* poll.[17]

When Theresa May became Conservative leader, more than forty years after Thatcher, she still had to battle a lot of sexism. There was the notorious 'Never mind Brexit, who won Legs-it!' front page of the *Daily Mail*, illustrated by a photo of May and the Scottish First Minister, Nicola Sturgeon, sitting next to each other, with the focus on their two pairs of legs. Inside, a column by Sarah Vine was headlined 'Finest weapons at their command? Those pins!' She referred to Sturgeon's legs as 'altogether more flirty, tantalisingly crossed . . . a direct attempt at seduction'. Really? They were discussing the future relationship between Britain and Europe, not the merits of a local lesbian nightclub.

In her book with Sylvia Shaw, *Gender, Power and Political Speech*, Deborah Cameron analysed the media coverage of female politicians in the 2015 UK general election. She wrote afterwards, 'It was noticeable that many of the most overtly hostile examples were produced by right-wing female columnists . . . These women are "Pets": they're rewarded for acting as mouthpieces for the prejudices of the men who control the Tory press. Their editors know that if a man described Nicola Sturgeon as a power-crazed Lady Macbeth with a haircut like a Tunnock's Teacake (I take this childish insult directly from a 2015 column by Allison Pearson), he'd come across as a crude chauvinist bully. So the task of trashing women gets delegated to the ladies, producing a steady stream of female-authored "why I can't stand [insert name of female politician]" pieces.'[18]

You might think that the press's treatment of female leaders would have improved over the past forty years since Thatcher came to power. Depressingly, the opposite is the case. A study by Blair Williams of newspaper coverage of both Thatcher's and May's first three weeks in office finds that the coverage of May was much more gendered than that of Thatcher, particularly in conservative newspapers.[19] May had twice as much written about her appearance and twice as much about her femininity. It is hard to come across as authoritative and powerful if people are constantly obsessing about your hair, your shoes or your handbag. This incessant critique of female politicians' appearance serves to undermine their authority.

All this, though, was when May was still riding high as prime minister, before she had to take on her party over Brexit. Then she faced stark challenges to her authority: she was openly despised by many of her own MPs, suffered disloyalty even from her Cabinet, and was horribly disparaged in the media. The trouble is, at least some of this was deserved. She was never cut out to be prime minister. Although she faced the near-impossible task of taking the UK out of the European Union without a parliamentary majority, it was she who had lost that majority by holding an unnecessary election in 2017. And she had very poor people skills and low emotional intelligence. She was as bad at dealing with her colleagues as she was with EU leaders and negotiators. It is hard, therefore, to disentangle how much of the criticism of her was sexist and how much was legitimate.

When male leaders fail, however, it is never taken as a reflection on their entire gender. No one said, after the disastrous Iain Duncan Smith was forced out as Conservative leader, that the party shouldn't risk another man. But when May looked as if she was going to face a leadership contest, Amber Rudd, who was then in May's Cabinet as Work and Pensions Secretary, had this exchange with a fellow MP. 'Just so you know, Amber,' he told her, 'if there were a leadership contest, I would want to back you. But I think we've had enough women for now.'[20] In his eyes, one woman failing meant that no other politician of the same sex should be allowed a chance. Half the population was automatically ruled out of the race.

If women do make it to the top of politics and can prove their ability, though, they can become very popular. Sturgeon has had extraordinary electoral success in Scotland. Angela Merkel has had her ups and downs, but is still seen in Germany as the mother – *Mutti* – of the nation and, at the time of writing, has spent fifteen years in power. Jacinda Ardern in New Zealand, like many of her female counterparts around the world, was seen to rise magnificently to the coronavirus crisis and subsequently won a landslide election victory.

And it's not just in their response to a global pandemic that female political leaders have excelled. A worldwide study by the Global Institute for Women's Leadership finds that women make rather good politicians all round.[21] On average, they do more constituency work than men. They tend to be less corrupt. Their leadership style is more cooperative and inclusive. They bring into politics issues such as domestic violence and female genital mutilation and put more energy into areas such as family and childcare, reproductive health and equal rights. They prioritize education, health and welfare, while spending less on defence and more on aid. All in all, the report says, 'More women leaders seem to make for more equal and caring societies.'

They also seem to be better at fostering economic growth in countries that have conflict between ethnic groups. A study of 188 countries over more than half a century found that, in countries with the greatest ethnic diversity, women leaders produced an average of 5.4 per cent annual GDP growth, compared with 1.1 per cent for male leaders.[22] Of course, correlation doesn't guarantee causation, but the authors write that 'there is reason to believe that these female heads of state actually led their diverse countries differently than their male counterparts'.

There is also a pattern of female political leaders using a more humble, democratic style that appeals to voters. Corazon Aquino, President of the Philippines, declined to live in the presidential palace but instead worked out of a small office. Golda Meir, Prime Minister of Israel, insisted that committee meetings were conducted without hierarchies – 'like a kibbutz'. The Irish president Mary Robinson kept the door of her house open to the people of Ireland.

When Jadranka Kosor took over as Prime Minister of Croatia, the country was in the throes of an economic crisis. So she decided to lead by example, to the horror of some of her colleagues, she told me. 'I cut some amenities, such as free coffee and free meals, which also included free meals on the government plane. During a trip to Brussels, my colleagues asked me, "Why did you do this?" And I responded, "Well, we can't ask Croatian people to live so austerely if we don't set an example ourselves."'

They complained that they were hungry. 'So on the next trip, I stopped on the way to the airport and bought a bag of pastries at the bakery and gave them to my colleagues on the plane. Later on, they would buy pastries themselves and would say, only half-jokingly, "Our boss is so strict she doesn't even let us eat!" '

Perhaps it isn't surprising that women leaders tend to be more successful, as women have to be better than men to be elected in the first place. Although, on average, women fare as well as men electorally when they run for office, if you control for characteristics such as competence and integrity, women are at a 3 per cent disadvantage.[23] In other words, the women who *are* elected are, on average, more competent and honest than their male counterparts. Women still have to straddle the authority gap in order to get elected, which means they usually have to be better than their male rivals. And the reason they have to be better is that men are otherwise reluctant to vote for them, preferring male candidates.

It was men's reluctance to countenance a female president that led to Hillary Clinton's defeat in 2016. Women, even white women, voted for her just as strongly as they had voted for Obama in 2012. But too many men who had supported Obama switched to Trump: enough to cost her the presidency in what was a very close race.

Maybe what these men need is to see the example of a successful woman in office. A study in Mexico found that in cities with a female mayor, men were much less likely to agree with the statement that 'men make better political leaders than women'.[24] It may turn out, in the US, that the first female president wins the position by default. Given Biden's advanced age, there is at least a chance that Kamala Harris will take over as president halfway through a term. Then she would be able to prove herself in office before facing the electorate again. And voters, by then, might be less fearful about the prospect of a female president.

If Harris can successfully tread the perilously narrow path between being assertive and competent enough to lead but warm and communal enough not to evoke too much dislike, then she will have a chance of becoming America's first elected female president in 2024. And if she succeeds in that, she will make it so much easier for the women who follow her – and, indeed, easier for women who want to wield authority all over the world in other walks of life.

For it is having women routinely in positions of power, in every organization, that will, in the end, abolish the authority gap. We will realize that the sky doesn't fall in when women run things. Once it no longer becomes unusual for women to be in charge, we will feel much less discomfort at the incongruity of it, just as we no longer experience a jolt of surprise when we see a woman driving a car, though I bet my grandparents did in the 1920s.

Of course, this is chicken-and-egg: the authority gap makes it harder for women to get to the top in the first place. But more women *are* winning power, and their success will start to eat away at our stereotypes, which in turn will reduce our unconscious bias, enabling more women to follow them. It may be a slow process, but we can choose to speed it up. I'll explain how in Chapter 15.

12

Bias entangled
The busy intersection of prejudice

'AND YOU MUST BE PUSHY'

'We're "leaning in" so far we're flat on our faces. Even if I keep leaning in, I need someone there to open the door.'

– Katherine Phillips, professor at
Columbia Business School

'Of all groups, as bona fide intellectuals, African American women are the furthest removed from society's expectations of their place, the least expected to succeed on merit, and the most vulnerable to insult.'

– Nellie Y. McKay, professor and author

'I WAS ALWAYS THE first in my class, the first in my form, and the class representative, but when I was fifteen and had to go to see a psychologist to assess whether I should stay on at school, she told me to leave and do sewing. I found out later on that it was not at all unusual; it was systematic. I talked to so many black people who had the same story. Was it my colour or my gender? Probably both.'[1]

Thankfully, Olivette Otele ignored the psychologist at her French school, went to the Sorbonne and ended up as the first black woman to be a history professor in the UK. She is Professor of the History of Slavery at Bristol University and was all over the airwaves after the killing of George Floyd and the resurgence of the Black Lives Matter movement. But still, she says, 'I have to prove I'm twice as competent as a white man to get on in life.'[2]

Margaret Casely-Hayford recognizes this. She is an enormously talented lawyer, leader and businesswoman. She's Chancellor of Coventry University, Chair of Shakespeare's Globe and used to be company secretary and head of legal affairs at John Lewis. In 2014, she was voted Black British Business Person of the Year. But she has met this problem all through her life.

'The first time I saw the authority gap in all its brazen irritating glory,' she told me, 'was when I was working in a legal department and a trainee solicitor came to work in the department. I was asked to give him some guidance on conveyancing, so I said, "Why don't you sit with me while I do this completion?" I went through everything and he sat and watched. Afterwards, the solicitor said to him, "I think she did very well, don't you?" The guy didn't say, "Well, actually *she's* training *me.*" I was the young black girl, so I had to be the trainee.

'People always think down to black people,' she says. 'The authority gap is much bigger in terms of white-to-black than male-to-female. It's absolutely enormous. That's the hardest thing to transcend.'[3]

Mamokgethi Phakeng agrees. Vehement and charismatic, she runs the University of Cape Town, Africa's highest-ranking university. 'As I went higher,' she told me, 'the fact that I'm black was much more of an issue than the fact that I'm a woman. The first disadvantage I experienced and I was conscious about was the disadvantage of being black. The assumptions that people make about you, what you can do, how capable or not capable you are; those kind of things came first, before gender or any other thing.'[4]

To see how race trumps gender when it comes to the authority gap, you only have to consider this experience relayed to me by Bernardine Evaristo. 'I remember being in a restaurant with one of my students, we were having a tutorial outside of the university and she was twenty-one, a young white girl. And the waitress came up and asked her for the order, and then gave *her* the bill. Now, I'm a woman in my fifties at that stage, but I was a black woman, so I was invisible to her.'[5]

As a white, middle-class, heterosexual, able-bodied woman, I am hugely aware of my limitations when it comes to writing this chapter. So I'm going to call on the experiences of many of the women I've interviewed who've battled much more than sexism in their lives to illustrate how complex the other biases are when they are entangled with gender.

I've mentioned the intersections between race and gender several times already, but in order to understand the insidious ways in which the authority gap manifests, a much deeper dive is required, especially when you add other factors, such as class, into the mix. Being black and working class makes it much harder to be taken seriously as a woman, as Evaristo has found. 'We know that gender is an issue, but race is an issue and the two are intertwined. And the fact that I come from a working-class background is absolutely part of it.

'The people who have positions of power in our society have for most of my life, and most of British history, been white, upper-class men. The government is more diverse than it has been in its history, but it's still basically run by white men, often coming from very privileged backgrounds. And so that's the default when people think about people in authority. So if you are coming from a background where you are black or Asian, you're a woman or you're working class, then I think you have a battle on your hands, because people don't automatically think that you should be in a position of power.'[6]

Conversely, being upper class can shield some women. So, for instance, the novelist Kamila Shamsie told me that her high-class status protects her against a lot of sexism when she is in Pakistan, because it is such a class-conscious country. 'In Pakistan, my class privilege is immense. I mean, it enters a room ten paces ahead of me. I'm not saying that class privilege means that the patriarchy doesn't exist. But it makes you operate in certain spheres of extreme privilege.'[7]

So what does the research tell us about the size of the authority gap for women of colour, working-class women and other minorities? Well, the title of a huge tome about the intersections of race and class for women in academia says it all: *Presumed Incompetent*. Each of these differences from the middle-class white male default widens the authority gap, though often in different ways for different ethnicities. But one way or another, the assumption is that these women will be much less competent than white men and they suffer treatment ranging from underestimation and disrespect to outright hostility.

As fifty-year-old Otele told me, 'Although I look my age, some people think sometimes that I'm a mature student. At a conference, I'd meet someone my equal and they'd tell me how brave I was, going back to study. Nowhere at any time did I tell them that I'd gone back to study; they just assumed because I was with a bunch of students that I was a student. I had to inform them that I was their lecturer. It's incredibly patronizing.'[8]

Laura Bates has come across this a lot: 'I can't tell you how many times we've heard [in the Everyday Sexism Project] from a black woman who is, for example, at a conference where she's due to give the keynote speech, and while waiting to go in, has white male attendees constantly coming up to her, asking her to direct them to the toilets, making the assumption that she works there. That's something that we hear over and over again.'[9]

Presumed Incompetent is full of stories by female academics of colour about how they have struggled to be taken as seriously as men or as white women, when racist stereotypes are overlaid on sexist ones. Sherree Wilson is associate vice-chancellor at Washington University School of Medicine. As she explains, 'Since some white students can only imagine African-American women as servants or caretakers, they may be unable to accept or adjust to the idea of having to work for African-American women. Those students may resent an African-American woman who has power over them and is in a position of authority; the result may be lower-than-average student evaluations and an increased number of complaints regarding assignments and overall teaching competence.'[10]

African-American women faculty often describe instances of having their credentials questioned or challenged by students. This resonated with a senior African-American academic whom Wilson calls 'Professor Andra': 'You're challenged about things like, "How do you know that?" or there is more testing and sort of questioning your right to give them anything less than an A.'

Andra struggled, too, to get her voice heard by her colleagues.

'There was a faculty meeting, and I was making comments which would be pretty much ignored. A white male would then make the same comment, and then everybody heard it – oh, isn't that brilliant. Those kinds of things would happen quite frequently. You're not heard, and myself and two other African-American women faculty colleagues, we made the most noise, but people would literally not hear what we said, and then we would have a white faculty member repeat it, and they would react as if it was the second coming.'

Did this result from race or gender? 'I think when people look at us, they see a black woman, not a woman who is black, and so the first thing that they relate to is your race.

'Your mistakes are noticed, very much so. And whatever short-comings you have, they're very, very visible. It's just when you get to your achievements, somehow you become less visible, but any shortcoming you have is magnified to the nth degree.' Black women have to outperform even more than white women to get a seat at the table, but even once they are there, they can too often be over-looked.

I quoted Professor John Dovidio from Yale earlier on in the book. He's the psychologist who, along with Professor Mahzarin Banaji, developed the Implicit Association Test for unconscious bias. As he says, 'The frequent questioning of competence that is directed at women of colour, which may sometimes be echoed by women and men of colour themselves, is not new. In the past, their competence was openly questioned, but now it is whispered about or held in silent suspicion. But this is not social progress. Research in psychology, sociology and political science demonstrates signifi-cant declines in overt expressions of racism, sexism and other forms of -isms as egalitarian principles become more widely endorsed. This trend does not mean that bias is disappearing, however; it is being replaced by prejudice and discrimination that may be less conscious and intentional and manifested in more subtle ways.'[11]

And this bias can even be held by minority ethnic women themselves. As Banaji herself told me, talking about her own experience of taking the Implicit Association Test, 'When I come face to face with the fact that I cannot associate dark-skinned people with good things as quickly as I can associate light-skinned people with good things, that's very different from just awareness. That's like someone putting a little dagger into me and turning it and asking me to sit up and take notice.'

These biases – both gender and race – feed through into career progression, explains Dovidio. 'Our research and the work of others show that black, Asian and white women who have impeccable qualifications may be hired or promoted at rates comparable to those of white men, but when their record is anything short of perfect, they are victimized by discrimination. In these cases, decision-makers weigh the strongest credential of white men most heavily while they systematically shift their standards and focus on the weakest aspects of racial minorities. The process often occurs unconsciously, even among people who believe that they are not racist or sexist. Moreover, because people justify their decisions on the basis of something other than race or sex – how a particular aspect of the record falls short of the standards, for example – they fail to understand the way racism or sexism operated indirectly to shape the qualities they valued or devalued and, ultimately, what they decided.'

And there's more. 'Beyond the biases and obstacles that women of colour face for being both women and people of colour, they confront an additional challenge: being invisible. Members of minority groups are perceived primarily through standards exemplified by the men of their group, and women of colour are typically judged by standards that are tailored to white women. It is thus more difficult to understand what a woman of colour means psychologically; she often "falls between the cracks". As a consequence, what she says and does is more easily overlooked or forgotten.'

As he says, it's more complicated than just adding together the discrimination caused by race and by gender. Adrien Katherine Wing, who teaches law at the University of Iowa, writes: 'To me that discrimination was multiplicative, not additive. In other words, I am Black times a woman every day, not Black plus a woman, which implies you may be able to subtract an identity. The discrimination I felt was against me as a holistic *Black woman*.'[12]

These intersections between racism and sexism can be very painful for those at the receiving end. Linda Trinh Võ, an American of Vietnamese heritage, is Professor of Asian American Studies at the University of California, Irvine. She says: 'I had to prepare myself mentally for the hostile teaching environment I faced every time I walked into the lecture hall. Those who were antagonistic made their opposition clear from their body posture to the blatantly racist comments they made in class, greeted by other students clapping and cheering in agreement. I was the first Asian American many of these students had ever encountered, especially a female in a position of authority.'[13]

Maggie Aderin-Pocock is a black British space scientist. 'When I was younger,' she recalls, 'when I was studying for my Ph.D., I was having lunch, a guy had come over from Holland, he was studying his Ph.D. there. And we'd had a nice lunch and just at the end of the lunch, he turned to me and said, "Whose secretary are you?" Everybody else on the table was studying for the Ph.D., but he assumed that I was somebody's secretary.'[14]

This bias is widespread. The presumption when a minority professor walks in the door is that he or she is not well enough qualified.[15] Black women faculty – even more than white women – report that white students often question both their competence and their authority. JoAnn Miller and Marilyn Chamberlin, in a paper called 'Women are Teachers, Men are Professors', found that students consistently underestimated the educational credentials and academic rank of women and minority professors.[16]

The bias faced by women faculty of colour comes almost entirely from white male students, and often their behaviour is rude and intimidating. Here's an account from Alice, who is black: 'White males will open my door to my office without knocking . . . No one else just opens up my door. They're snide, they'll sit with their arms crossed and they doodle and they sit right up in the front so that is definite passive-aggressive behaviour. The tone sometimes in the emails they send, it's the kind of things you don't even know how to express to other people. But I know if I was a white male they wouldn't dare write to me in that tone.'[17]

It's not just hostility from students that these women have to navigate, but from their colleagues too. They are excluded from networks by white men based on race and gender, by white women based on race, and by men of colour based on gender. In most organizations, not just universities, there aren't enough women of colour, particularly in senior positions, to help each other along.

Sometimes women of colour can form alliances with white women, but it doesn't always work out, as white women, even if they are allies on gender, benefit from a race privilege that women of colour don't have. Mamokgethi Phakeng explained it to me. 'There were times, many times, where I would be on the same side with white women. I was very fortunate to start my academic career at the University of the Witwatersrand because there were strong women who were all white, and I became friends with them. But you form an alliance with them and then there's a time where you see they get access to certain things and you don't, they get treated differently and you don't, so there were times when I would have confrontations with them because their whiteness got them a ticket that I couldn't get. And then there were times when we would be on the same side because the men would be getting tickets that we don't get. So it's been a journey of dealing with race and then gender, at some stage being with the white women allies and sometimes being on your own because there aren't many people of my race and gender in the discipline.'[18] It is vital for white women (and men) to act as allies to female colleagues of colour, just as women need men to act as allies if we are to narrow the authority gap more generally.

Olivette Otele found outright hostility from her colleagues when she was promoted to professor. 'I didn't receive congratulations from any of the junior men when I was appointed,' she told me. 'The initial reaction was surprise and annoyance. The reaction of men my age and older was nasty. They didn't think I was qualified for promotion. Some sent messages saying they thought it was a bit early for me to be promoted. I said to them, "Did you bring £1.6 million to the institution? I have!" That's the range of grants I've won over the years. "Just go crawl back to your cave," was what I wanted to say. They basically thought it was political correctness gone mad.'

'Political correctness gone mad' or, in American parlance, 'affirmative action hire', is the accusation often levelled at women of colour who are hired or promoted. It's true that hiring a woman of colour is a good move for organizations keen to improve their diversity, but that doesn't mean that these women are any less deserving of, or qualified for, their jobs. Unfortunately, though, the accusation haunts them through their careers, gives their colleagues licence to disrespect them and can be hugely undermining of their confidence.

As Yolanda Flores Neimann writes: 'There is strong documentation for the idea that a stigma of incompetence arises from the affirmative action label, especially when the label carries a negative connotation in the hiring department. Once tagged as an affirmative action hire, colleagues may discount the qualifications of the hire and assume [the candidate] was selected primarily because of her minority status, thus leading to the presumption and stigma of incompetence.'[19]

And that's if the women of colour get hired or promoted in the first place. Black women are more likely than white women (44 per cent to 30 per cent) to report feeling stalled in their careers, and to feel that their talents aren't recognized by their superiors (26 per cent to 17 per cent).[20]

Each year, the consultants McKinsey and Lean In do a big study on 'Women in the Workplace'.[21] In 2020, they surveyed more than 40,000 employees. In the C-suite – the jobs at the very top of organizations – only 3 per cent were women of colour, compared with a still pretty meagre 19 per cent white women.

It's not as if black women are any less ambitious; if anything, they are quite a lot more so. According to a recent Nielsen survey, 64 per cent of black women in the United States agree that their goal is to make it to the top of their profession; that's nearly double the percentage of non-Hispanic white women with the same goal.[22] And black women are far more confident than white women that they can succeed in a position of power (43 per cent to 30 per cent).

But black women are not ascending to top positions in any significant numbers, and looking at their answers to the Women in the Workplace researchers' questions, you get a measure of why. Only 42 per cent of black women as against 57 per cent of all women say they have an equal opportunity for advancement. Only 35 per cent (as against 48 per cent) say promotions are fair and objective. And just 29 per cent (as against 37 per cent) say their manager advocates for new opportunities for them. In all these categories, Asian and Latina women say they do better than black women but worse than white ones. As a black female senior manager told the researchers: 'A lot of Black women think that many gender initiatives are really tailored toward white women. Are they targeting women of colour too? It often doesn't feel like it.'[23]

This is partly because they are often excluded from the networking and mentoring opportunities that white men in particular, but also to some extent white women, benefit from. So, for instance, white men report having access to senior leaders in their organization at three times the rate of black women, and white women do at twice the rate. Only 19 per cent of black women say they have had a mentor or sponsor in their career, compared with 30 per cent of white women.[24]

Meanwhile, the evidence of the authority gap is bleak. Forty per cent of black women say they are asked to provide more evidence of their competence, compared to 28 per cent of white women and 14 per cent of men. And 26 per cent of black women report hearing others' surprise at their language skills or other abilities, compared to 11 per cent of white women and 8 per cent of men.

A study of women leaving STEM jobs found exactly this problem. Two thirds of the women interviewed reported having to prove themselves over and over again – their successes discounted, their expertise questioned. ' "People just assume you're not going to be able to cut it," a statistician told us, in a typical comment. Black women were considerably more likely than other women to report having to deal with this type of bias. (And few Asian-American women felt that the stereotype of Asian-Americans as good at science helped them; that stereotype may well chiefly benefit Asian-American men.)'[25]

In 2016, a Facebook post by the black doctor Tamika Cross went viral. She told the story of how she was prevented from helping a sick passenger because the flight attendant couldn't believe that she was a medic. 'Oh no, sweetie,' said the attendant, 'put your hand down; we are looking for actual physicians or nurses or some type of medical personnel. We don't have time to talk to you.' Dr Cross wrote, 'I'm sure many of my fellow young, corporate America working women of color can all understand my frustration when I say I'm sick of being disrespected.' After her story surfaced, female doctors of colour queued up to tell the world how they, too, had been written off in the same way.

Kadijah Ray, an anaesthesiologist, wrote: 'I've received that same treatment on two different flights in 2006 and 2008 while trying to help people in distress. They passed me up for whites: a female pharmacist, a nurse and a male MD who I believe was in something like radiology. I remember him telling them, "Trust me, you want an anaesthesiologist to help before me." And no, I didn't have my credentials with me to prove I was qualified. They would far exceed the airline's weight and size requirements.'[26]

Ashley Denmark, a black doctor, was turned away by flight attendants in favour of two white nurses, even though she showed them her hospital badge. 'The gravity of the situation hit me like a ton of bricks,' she said. 'Apparently the nurses and flight attendants didn't think I was a doctor. Why else were nurses being allowed to take charge in a medical situation when a doctor was present? Surely it couldn't be the colour of my brown skin? So here I was, a doctor with eleven years of training, being asked to take a seat and not partake in caring for the passenger in need.

'As an African American female physician, I am too familiar with this scenario. Despite excelling academically and obtaining the title of "doctor" in front of my name, I still get side-eye glances when I introduce myself as Dr Denmark. Commonly, I'm mistaken for an assistant, janitor, secretary, nurse or student, even when I have my white coat on.'[27]

An American study called 'Who Benefits from the White Coat?' finds that female Indian doctors have a much harder time than male ones.[28] Because the profession of doctor is held in such high esteem in the US, male Indian doctors are instantly respected and accepted as soon as they tell people what they do. For women of the same background it is quite different, says the author Lata Murti: 'The female doctors said that their membership in the masculinized profession of medicine was often questioned in public, even while wearing their white coats and hospital identification badges. The

232

stereotype that doctors are men is so prevalent in non-clinical spheres that Americans of all races have trouble conceptualizing women as doctors. Add to this the stereotype that non-white immigrant women are submissive, financially dependent and limited to traditionally female roles, and the idea of a brown woman in a white coat has no place in the American imagination.

'When brown Asian women like the female interviewees have an occupational status higher than many white men, they defy Americans' gendered racial expectations. They become socially undesirable because they are perceived to have achieved professional equality with white men without "passing", or suppressing their negatively racialized feminine traits. The female Asian-Indian doctor represents a type of pariah femininity in the USA, namely the aggressive, authoritative "bitch". Unlike the male doctors, they risk losing social acceptance and desirability whenever they reveal their occupation.'

Here's the first-hand experience of Deepti, a female Indian doctor: 'Whenever I start a new job, or when I was training for my residency and fellowship, I am incredibly aware that being a doctor is seen as a man's profession. So number one, being a female you have to jump through extra hoops to establish yourself. And then, when you look like me and when you talk like me you stand out . . . if you are a new member of the team nobody is going to accept you . . . I have to prove myself, right? I have to prove that I am capable of taking care of the twenty patients that I am supposed to take care of. So when you are a female physician, and then you are a female "not Caucasian", then you are definitely made to prove yourself twice over, before they accept you as one of them.'[29]

The role of stereotypes is as strong for race as it is for gender. Above we saw the stereotype of the demure, submissive Asian female. Yet for black women, it's very different. A study that asked people to come up with ten characteristics for different racial groups and genders found that 'confident', 'assertive' and 'aggressive' were all in the top fifteen characteristics given for black women. None of these adjectives made it into the top fifteen for white, Latina, Middle Eastern or Asian women.[30]

This plays into the damaging stereotype of the angry black woman. So many women of colour told me how hard it was to avoid this. Here's Anita Martin, a psychiatrist in the north of England. 'When I was a junior doctor, I spoke up for my colleagues on a health and safety issue. That went down like a lead balloon. It was a constant, constant thing all the way through. How dare I have an opinion and act like a white person? If you're coloured and you're female and you're assertive, you're an angry, uppity black woman. Your average white man gets very cross because I don't see myself as lesser.'[31]

Bernardine Evaristo thinks this stereotype is a deliberate ploy to keep black women in their place. 'It's a way to keep us passive and docile. And to strip us of our power. It's very toxic.'[32]

'Black and Latina women are particularly at risk for being seen as angry,' writes Joan C. Williams, author of a study on the biases that drive women out of STEM careers.[33] 'A biologist noted that she tends to speak her mind very directly, as do her male colleagues. But after her department chair angrily told her, "Don't talk to me like that," she now does a lot of deferring, framing her requests as, "I can't do this without your help."' She explains, 'I had to put him in that masculine, "I'll take care of it" role and I had to take the feminine "I need you to help me, I need to be saved" role.'

Interestingly, though, and in keeping with the finding that black women are allowed to be more agentic, only 8 per cent of the black women in her study agreed that, at work, they find themselves pressured to play a stereotypically feminine role – way lower than the 41 per cent of Asian women and 36 per cent of white women who did. And only 8 per cent of black women said that colleagues suggested they should work fewer hours after they had children, compared with 37 per cent of Asian women.

In another study of stereotypes, Asian women were twice as likely as black or white women to be thought to be intelligent.[34] They were ten times more likely than black and white women to be thought 'mild-tempered', and three times more likely to be 'subservient'. Where black women scored more highly than Asian or white women was in anger, strength, dominance, achievement-orientation and being interesting. But if black women are allowed to be stronger and more decisive than women of other ethnicities, when they fail, they are judged more harshly than white men, white women and black men.[35]

'Why shouldn't I be angry?' Mamokgethi Phakeng asked me. She told me how she had grown up during apartheid in a country that had been colonized. She was born into poverty but managed to get the education that her father told her would be a ticket out of it. She climbed every rung of the academic ladder. 'I wanted to make sure that I tick all the boxes because I thought if you do this, you will escape racism. Now you get there, you're here and then you still don't win. Now they can't find fault with my work, so they tell me I'm a narcissist, I'm a bully. Men who don't have doctorates become vice-chancellors of top universities and they're not critiqued in the same way.

'Why wouldn't you be angry? Why? Tell me why? The fact that I can still sit in a room and calmly have conversations with white people and have white friends and hug them and have collaborations with them, the world should sit up and say, "These black people are amazing!" So nobody should be surprised that you have such anger, they should be surprised that we've been so forgiving. Why shouldn't black women be angry when they are still at the end of what I call the colonial procession? Why? We should be *more* angry. In fact, we are very peaceful.'[36]

*

Phakeng grew up poor, and – as Bernardine Evaristo pointed out earlier – the intersections of class and race are powerful, and all too often overlooked. Constance G. Anthony is a black political scientist at Seattle University. She writes, 'My journey through the academy as a gay, working-class woman in an overwhelmingly straight, middle- and upper-middle-class male field has been constrained by each of these social states, but despite the intersectionality of these pieces of identity, class is the least socially recognized and, perhaps for that reason, the most corrosive.'[37]

She has a point. In Britain at least, as soon as someone opens their mouth and begins to talk, we tend to judge them by their accent, which can be such a class signifier. A study by the Social Mobility and Child Poverty Commission found evidence that recruiters favoured people with some accents over others, regardless of their academic achievement.[38] And what they were particularly looking for was 'polish', which is a euphemistic way of saying 'middle class'. Although the prejudice in favour of received pronunciation is weakening, one interviewee told the researchers, 'In my first appraisal with my then partner, he made a comment to me that because I was from the north of England I had to be very careful that people didn't think I was a . . . fool.'

Lance Workman and Hayley-Jane Smith decided to put this to the test. They asked people to rate how intelligent a young woman was when reading a passage in a Yorkshire, Birmingham or received pronunciation accent. They also added a silent option, where people saw the face but didn't hear the woman speak. In order of intelligence, people rated the Yorkshire accent top, followed by RP. The Birmingham accent came below the silent one.

This suggests that RP may no longer be top dog, perhaps as a result of so many more people with regional accents going to university. Yorkshire accents are perceived as more trustworthy, and trustworthiness is associated with intelligence. As for the Birmingham accent, it's how the co-author of the paper, Hayley-Jane Smith, talks and, according to her colleague Dr Workman, she's 'extremely intelligent'.

Similar research done in the US found that women with a Spanish accent were thought to be less knowledgeable than women with a North American one, particularly by men.[39] In almost all cultures, upper- and middle-class people are perceived as more competent but cold, while working-class people are seen as less competent but warm.[40] The perceived incompetence of poorer people is even worse in highly unequal societies.

Part of this may be bound up with confidence. A study conducted in Mexico and the US found a strong correlation between high social class and overconfidence.[41] And this overconfidence (as we saw earlier in the book) led recruiters to think that those of a high social class were more competent. Because of their double disadvantage, working-class women are even less likely to be over-confident.

The writer Bel Mooney comes from a working-class background, and it made her feel as if she didn't belong. 'I've been aware of this all my adult life,' she told me, 'from student seminars through my first marriage into my whole career in journalism and broadcasting. With me it was a mixture of class and gender holding me back. The class consciousness (at the beginning) my very own problem; the mansplaining arrogance all through *their* problem.'[42]

Working-class women really understand how hard they have to work to be recognized as equals. Cherie Booth, a fiercely intelligent human rights lawyer (and, as it happens, wife of Tony Blair), was brought up by a working-class single mother and grandmother in Liverpool. 'When I went to university, less than 10 per cent of my class were women in the law, because it wasn't something that girls really did,' she told me.[43] 'But I came top of the first-year exams, the second-year exams and the third-year exams, partly because I knew that if I didn't do well, there was no chance in hell of me getting a job. The only way I could justify my being there was because I was cleverer than everybody else. That was a class thing, not a girl thing. Everyone else in my family had been working by the time they were sixteen.'

But it wasn't long before it became a gendered thing too. 'I came top of the Bar finals but the Bar is full of people who are the best, and it's certainly full of people who *think* they're the best! And it was certainly full of people who thought men were the best, and that's when suddenly I was told that because I was a woman, clearly I didn't have the authority. I'd go to court and people thought I was the secretary. When it came to chambers choosing between me and Tony, there was the boy from public school and Oxford with a 2:1 and a 3rd in his Bar finals or the working-class girl from Liverpool with a First and top of the Bar finals and obviously they chose him. So at that point I suddenly thought, "Oh my God, there's a problem here because I'm a girl." '

We've looked at the intersections of race and class, but what about sexuality? Are non-hetero women doubly disadvantaged when it comes to authority? Interestingly, the picture is more complicated.

The McKinsey / Lean In 'Women in the Workplace 2019' report, which surveyed more than 38,500 employees, finds a very mixed experience for lesbians. On general career statements, such as 'I have equal opportunity for growth and development'; 'I have equal access to sponsorship'; or 'My manager gives me opportunities to manage people and projects,' lesbians score more highly than straight women, and are almost as positive as men.

Yet when it comes to everyday experiences of the authority gap, lesbians report worse treatment. They are more likely than straight women to say they need to provide more evidence of their competence, that they have their judgement questioned in their area of expertise, that they are interrupted or spoken over, and that others take or get credit for their ideas. Less surprisingly, perhaps, they are also much more likely to say that they hear demeaning remarks about them or people like them and that they feel they can't talk about themselves or their lives outside work.

In some ways, lesbians tend to do better in the workforce than straight women. On average, lesbians earn more than heterosexual women, whether or not they have children.[44] This is partly because they are more likely to work full-time, even if they are mothers. And they are more likely to have jobs in traditionally male fields. According to one study, 'Some found that being a lesbian was a distinct advantage in a male environment. They felt that they did not have to deal with the same amount of sexual advances and harassment as their female colleagues in clerical and secretarial jobs. They generally reported feeling very comfortable with their male co-workers and interacted with them as friends.'[45]

Lesbians tend to have to do less of the 'second shift' than straight women, because their partners are more likely to share the unpaid work at home equally. And they can often escape the feminine stereotypes that are used to hold other women back. Lesbians as a group tend to be rated more highly than heterosexual women by other people on independence, assertiveness, competitiveness and self-confidence – qualities that are useful for a woman at work.[46] Here's the experience of a lesbian Harvard Business School graduate: 'To most people, being a lesbian means you are focused on your career, not your husband and children, and you have a strong, aggressive style – just like other top executives.'[47]

So lesbians seem to occupy a rung on the ladder, when it comes to stereotypes, that is above straight women but below straight men and they do particularly well compared with heterosexual women when they start a family. In a study called 'The paradox of the lesbian worker', researchers asked people to assess the competence and career-orientation of a hypothetical man, a straight woman and a lesbian woman in a high-powered job (in this case, McKinsey consultant), both childless and as a new parent.[48] The perceived competence and career-orientation of the straight woman fell sharply if participants were told she had a baby. For men, by contrast, both factors rose after fatherhood. For lesbian women, competence rose and career-orientation stayed the same. So it looks as if lesbians – unlike straight women – don't suffer from the motherhood penalty: the assumption that motherhood makes them worse at, or less committed to, their jobs.

However, lesbians and bisexual women are more likely to be bullied and sexually harassed. The 'Women in the Workplace' study found that 62 per cent of bisexual women and 53 per cent of lesbians said they had experienced sexual harassment, compared with 41 per cent of all women. (Interestingly, the figure for women of colour was only 34 per cent.)

This may be a reason why lesbians tend to be more secretive at work about their sexuality than gay men. They have many fewer role models in openly lesbian public figures and often they are excluded from the public conversation about homosexuality. A book called *The G Quotient: Why Gay Executives are Excelling as Leaders . . . and What Every Manager Needs to Know* sets out seven leading principles, including a focus on inclusion and collaboration, but doesn't even mention lesbians.[49] As the author of a study on lesbians and leadership says, 'This reinforces the patriarchy by claiming these principles as the exclusive domain of gay men.'[50] Again, women are being seen as lesser, and not worthy of attention.

Bisexual women often feel that they are taken less seriously than lesbians. It is as if society expects women to choose either to be gay or straight and sees bisexuality as something ambiguous or indecisive rather than a valid choice. Beth Watson, 33, is the co-founder of Bechdel Theatre, an organization that campaigns for more diverse representation on stage. She is bisexual, and told me: 'One of the things that impacts us and doesn't affect gay and lesbian people is that our sexuality isn't taken seriously. We're told, "You're in a phase," or "It's a teenage thing when you're experimenting." There's often an experience of not being taken seriously that makes me hold back from telling people about my sexuality in a way that gay people don't. The sense of putting yourself out there as a publicly bisexual person. It's seen as not a fully formed thing, sitting in an unsure or uncertain space.'[51] And if women are seen as unsure or somehow immature, that lessens their authority and widens the gap.

The most disadvantaged group of all, though, are women with disabilities, who have 'been invisible, both to the advocates of women's rights and of disability rights', according to a background paper put to UN Women.[52] In the 'Women in the Workplace' study, they were the least likely of all categories to say that they had equal opportunities at work. They were also the least likely to say that their managers helped them to get on in their careers and some of the most likely to experience the everyday annoyances of the authority gap, such as having their expertise challenged and having others get or take credit for their ideas. Sadly, they are also the most likely to hear demeaning remarks about themselves or people like them.

And during the coronavirus pandemic, their lives were the hardest of all. Sixty-one per cent said they were stressed, 46 per cent exhausted and 40 per cent burned out: roughly ten percentage points more than the figures for all women.[53]

The average working woman with disabilities is paid only 83 per cent as much as a working man with a disability and 67 per cent as much as a working man without a disability. They also earn only 80 per cent of what able-bodied women are paid. Negative stereotypes play into this: people with disabilities – and particularly women – are often judged as weak, dependent or incapable by others.[54] As a result, many of them have to resort to working even harder than their colleagues to prove their capability. And we are talking large numbers here. In the UK, 19 per cent of women of working age are disabled.[55]

Emma Lewell-Buck, Labour MP and one of only five disabled members of Parliament, says that she used to take work home with her at weekends, work late into the evening and start early in the morning because, like many other disabled people, she felt she had to 'go the extra mile' and 'work that little bit harder to prove yourself or keep up'.[56]

Chloe, 27, who trains actors in improvisation, is both queer and disabled. 'That's a big part of my life,' she told me. 'My disability is almost worse than my gender sometimes. It's so personal to your experience that it can undermine any authority. I teach people older than me, so being young, female and disabled can be very difficult. Often people have a mistrust of what I know. I do a lot of research and people don't like it when I quote research at them.'[57]

Baroness (Sal) Brinton is President of the Liberal Democrats. Thanks to rheumatoid arthritis, she has spent the past decade in a wheelchair, so she has seen for herself how the authority gap widens for women once they become disabled. 'For some people, you really do become completely invisible,' she told me.[58] 'You just have to learn to be brazen, to break in, to get people to look at you, not to talk over your head, not to talk to your chief of staff because you're in a wheelchair and therefore you're dumb. It's hard work. Everything is hard work because your interpersonal relationships are completely different to people who can move around easily and look around easily.

'I'm a fairly forceful personality, so I make it very plain very quickly that I am on top of my game, I understand my brief, whatever. But the initial reaction is always to turn to the person they perceive as your carer. Always.'

Once this happened when she was being taken to the plane at an airport. 'Over my head, one member of staff said to the other member of staff, "Can she get herself to the gate?" And the member of staff who had brought me from security said, "I haven't a clue, perhaps you'd like to ask her? I believe she speaks."' Just think how hard it is to persuade people to accord you authority if they are not prepared even to address you directly.

Women who are neurodiverse often fall through the cracks. Autism and ADHD are diagnosed in four times as many boys as girls, but the actual prevalence may not be as skewed. Because there is so much more pressure on girls to be accommodating and social at an early age, that may disguise any challenging behaviour. Lily, 29, who has ADHD, told me, 'At school, I struggled terribly with timekeeping, keeping track of my assignments and getting my homework done on time. I'll never forget a teacher pulling me aside after yet another detention and telling me to "drop the ditsy girl act", as it wouldn't do me any favours. I am the farthest thing from ditsy. I am, in fact, to re-appropriate a sexist phrase, extremely ballsy. I just also happen to have a deficit in my executive function. Were the (largely neurotypical) boys in the detention called ditsy? Like hell they were.'[59] So ADHD boys were perceived as intelligent but with a medical disorder, while ADHD girls were seen as ditsy, or a bit dim. Lily, by the way, went on to read philosophy at Cambridge.

If it is exhausting enough for able-bodied, middle-class white women to fight the old gender stereotypes, it is multiply hard for women who find themselves up against other biases too. They have to prove their competence even more and, if they do succeed, they come up against assumptions that they did so only because of a box-ticking diversity exercise.

It is incumbent on those of us who have white privilege or class privilege to act as allies to women who don't. Obviously, the best method is to ask them how we can help. But we can deliberately mentor and sponsor more junior women of different races and backgrounds, and we can ensure that women of colour or lesbians, working-class or disabled women at work are included in our social networks. We can champion them to more senior colleagues and we can ask our employers to track the progress of women of different backgrounds separately, to make sure that some are not held back more than others.

Most of all, like the other biases that we have learned about, we need to acknowledge that, however liberal and compassionate we think we are, we are probably still unconsciously racist, homophobic, classist and ableist as well as sexist. We have to notice these biases as soon as they try to trick our brains and make sure that we correct for them in all our interactions with other people.

13

All things bright and beautiful

Or maybe if you're beautiful, you can't be bright?

'NO NEED TO BOTHER HER — I'VE GOT THIS '

'Math class is tough.' 'I love dressing up.'
'Do you want to braid my hair?'

– Teen-Talk Barbie

'Attack the Cobra Squad with heavy firepower!'
'When I give the orders, listen or get captured.'

– GI Joe

I WAS ONCE IN the audience at a big conference in London for investment companies when a man on stage asked why the *Financial Times* hadn't foreseen the global financial crisis. When it was pointed out to him that Gillian Tett, then assistant editor of the paper, had done exactly that, he replied, 'Oh well, she was too pretty for me to take her seriously.' I started a low 'Boo!' which – gratifyingly – spread around the hall.

From our earliest years, we absorb the notion that girls and women are designed to be ornamental and boys and men to be instrumental. As adults, we reinforce these stereotypes, often unwittingly, when we engage with children. 'What a pretty dress!' we might exclaim to a little girl. 'What a great footballer you are!' we might say to a little boy. To win approbation from the world, girls and women have to look good, but boys and men have to do well.

'We raise our boys to view bodies as tools to master the environment, and girls as projects to work on,' Caroline Heldman of Occidental College explained to me.[1] 'We take the yardstick of us grown women, our bodies, and give it to girls, saying, "Here, this is how you're supposed to value yourself," every time we say, "Oh, you're so pretty." '

And this can have damaging consequences for the girls in later life, she says. 'Over two decades we've done research finding that the more women see themselves as a sex object, the higher their eating disorders, the lower their cognitive functioning, the lower their happiness, their political efficacy [the idea that their voice matters in politics], they are more likely to be depressed, more likely to engage in habitual body monitoring. So we know that the more you think of yourself as a sex object, the more negative ramifications it has at an individual level. But we are taught way before we are conscious that our body is our primary form of value.'

This is the context in which women have to engage with the public sphere. Their appearance *matters*, much more than that of men. It has an effect on how seriously they are taken. It is constantly appraised and may be mercilessly critiqued. The authors of a study of how the British media cover politicians wrote that the judgements made about women were never applied to men; 'Male colleagues were to be found with lank and dirty hair, dandruff on their collars, stained ties, unsure about the precise positioning of their trouser waistbands (over or under their paunch) and their suits looking as if they had doubled as sleeping bags. If a woman were to appear in a similar state of dishevelment, she would make front-page news that day and questions would be asked about whether she was fit to be a Member of Parliament.'[2]

This is so true. I once wrote a column in *The Times* about the double standards of appearance that are applied to men and women in public life. I lamented the fact that a tiny patch of cellulite on Princess Diana's leg was all over the front pages, while the then Deputy Prime Minister, Michael Heseltine, *the second most powerful man in the country*, routinely had snowstorms of dandruff on his shoulders and nobody ever mentioned it. He was furious, and barely spoke to me again, but I did notice that the next time I saw him, his shoulders were pristine . . .

Not only is women's appearance minutely monitored, but there is a worrying trend to expect that successful professional women should also, at the same time, be sexy – a completely baffling conflation of categories. When the *Observer*, of all newspapers, ran a profile of Christine Lagarde, then the new cerebral head of the International Monetary Fund, it was headlined 'Is this the world's sexiest woman (and the most powerful?)'.[3] Note the order of the adjectives. Before the profile even cited the high-level jobs she had held that qualified her to run the IMF – French Minister of Finance, chair of an international law firm – the second paragraph read: 'What lovely teeth she has – straight and white, they gleam out of a permanently, almost alarmingly, tanned face. Tall – she's 5ft 10in – and slim, the 55-year-old Lagarde dresses with the casual *élan* of a Parisian, patriotically attired in Chanel suits and Hermès scarves, along with jazzy bracelets and fur-lined ponchos. Lagarde softens her rather severe black-and-white outfits with silk scarves, a string of pearls or a brooch. She has widely spaced green eyes framed by a silver bob.'

This was a serious, progressive paper. Yet here was a woman chosen for her intelligence, her leadership ability, her financial acumen and the respect with which she was held by international heads of government, and we were being asked to judge how sexy she was. Did she have good legs? A nice arse?

We wouldn't expect the Secretary-General of the UN, António Guterres, to look sexy, let alone be the sexiest man in the world. Or David Malpass, the President of the World Bank. It's not even as if Lagarde plays on her sexuality. Indeed, the same *Observer* profile quoted Andrew Hussey, a professor at the University of London Institute in Paris, as saying, 'She's unusual among French female politicians in that there's nothing coquettish about her.'

I've used the contrasting experiences of Heseltine and Lagarde to illustrate just what a perilous line women have to walk when it comes to their appearance, and how it affects how seriously they are taken. I've never particularly wanted to run the free world, so the only time I've felt seriously envious of Barack Obama was when he said he never had to decide what to wear in the morning because all his outfits were the same. Blue or grey suit, light shirt, boring tie. How simple life would be if all the stress and effort women felt compelled to put into their appearance could vanish – pfff! – like that.

Women have to think about so much when they are choosing what to wear. As the fabulous feminist writer Caitlin Moran puts it: 'Every morning, when a woman gets dressed, she is running potential outfits through a vector of factors before she makes a decision. Will these clothes make me look professional *and* thin *and* "nice" *and* "unique" *and* "with it" – *and keep me safe*? If you have ever wondered why women often say, "I have nothing to wear," despite having a wardrobe full of clothes, the answer is here: she means, "I have nothing to wear *for who I need to be and where I need to go today.*" '[4]

If you're a woman in public life, you have the added problem that if you wear the same outfit more than a few times (and sometimes, just twice), you are pilloried for it. As Angela Merkel discovered, 'For a man, it's no problem at all to wear a dark blue suit a hundred days in a row, but if I wear the same blazer four times within two weeks, the letters start pouring in.'[5]

Quite apart from the terrible waste, if women are expected to wear the same dress only once, how are they supposed to find the time and money to buy and wear completely different clothes every day? Add to that the amount of time, money and effort it takes to look groomed and *soignée* as a woman, when men simply have to take a shower, shave and brush their hair (or not, in the case of Boris Johnson), and you can see why women find this constant harping on their appearance frustrating. Think what they could do with that extra time, energy and money!

'I've never gotten used to how much effort it takes just to be a woman in the public eye,' Hillary Clinton wrote in her account of the 2016 presidential campaign, *What Happened*.[6] 'I once calculated how many hours I spent having my hair and makeup done during the campaign. It came to about 600 hours, or 25 days. I was so shocked, I checked the math twice.'

The attractiveness bar is also set much higher for women. I noticed this when I was chairing a revival of *The Brains Trust* on BBC2. Among the wonderful set of intellectual luminaries we had was the novelist A. S. Byatt. She was wise, erudite and thoughtful and a terrific addition to the panel. But watching it back, I was jolted every time the camera rested on her, because I realized how unused I was to see a not especially attractive older woman on screen. Yet at that time, John Sergeant – who, for all his talents, looked like the human cousin of a French bulldog – was on our TVs most days as the BBC's chief political correspondent.

Women are also, much more than men, expected to look younger than they are. As Susan Sontag wrote, in an essay called 'The Double Standard of Aging', '[For women], only one standard of female beauty is sanctioned: the girl. The great advantage men have is that our culture allows two standards of male beauty: the boy and the man. The beauty of a boy resembles the beauty of a girl. In both sexes it is a fragile kind of beauty and flourishes naturally only in the early part of the life-cycle. Happily, men are able to accept themselves under another standard of good looks – heavier, rougher, more thickly built. A man does not grieve when he loses the smooth, unlined, hairless skin of a boy. For he has only exchanged one form of attractiveness for another: the darker skin of a man's face, roughened by daily shaving, showing the marks of emotion and the normal lines of age.

'There is no equivalent of this second standard for women. The single standard of beauty for women dictates that they must go on having clear skin. Every wrinkle, every line, every gray hair, is a defeat. No wonder that no boy minds becoming a man, while even the passage from girlhood to early womanhood is experienced by many women as their downfall, for all women are trained to continue wanting to look like girls.'[7]

This has implications for how seriously women are taken. Girls are taken less seriously than middle-aged women, yet middle-aged women are expected to do all they can to look more like girls. As Elaine Chao put it to me, 'There is much more pressure on older women to look younger than they are than there is for men. Which is horrible. It's a paradox. On the one hand, as we get older, we actually get wiser, more assertive, and more able to occupy equal footing. On the other hand, our looks work against us.'[8] Mary Beard echoes this in *Women and Power*: 'Craggy or wrinkled faces signal mature wisdom in the case of a bloke, but "past-my-use-by-date" in the case of a woman.'[9] No wonder over 90 per cent of Botox users and 92 per cent of cosmetic surgery patients are female.[10]

As well as making efforts to look younger, women have to try to calibrate every morning, when they get dressed and made up, what effect their appearance that day will have on their authority. Will they be taken more or less seriously if they paint their nails? Wear lipstick? Have their hair up or down? Wear trousers or a dress?

Helle Thorning-Schmidt talked me through these calculations. She is still youthful-looking, blonde and beautiful: not a gratuitous description, but all relevant characteristics to bear in mind when you read what she says. 'In my late thirties, when I became leader of my party, I felt very young, and I immediately changed my look to look a little bit more serious. For women to look more serious, that means hair up, longer skirts, more suited up, and that was a role that I didn't particularly like, and I freed myself from it later, but it was something that I felt was necessary when I was so young and being the Leader of the Opposition.'[11]

But she still faced quite poisonous criticism for being apparently too glamorous. 'I was called "Gucci Helle" because the way I look doesn't fit well with how people think a politician should look. I never wore Gucci clothes, but I had a Gucci bag, that's where it came from, and I never regretted it either. I mean I like a bag, and I also think women have to keep being women in politics.

'But being called "Gucci Helle" didn't help my image as a political leader very much, especially leading a left-of-centre party. It implies that you are too posh for the party (which I never was, of course), that you're not a real social democrat, that you are interested in shallow things, that you don't have a proper brain for politics. So that is why it is a brilliant nickname for someone who wants to hurt you both inside your party and outside your party, and that's why it stuck so well.'

Michelle Bachelet, former president of Chile, has discussed this unfair criticism with many other female political leaders, and she told me that the overemphasis on their appearance is a deliberate attempt to delegitimize them. 'They tried to diminish the power of the women by criticizing things that are not substantial, like the size of their bags or the clothes they wore.'[12]

But if it's bad for elected female politicians, it can be even worse for wives of male ones. Cherie Booth is an extremely successful lawyer, but she also happens to be married to Tony Blair. So the transition to buttoned-up prime minister's wife was a particularly painful one for her, especially as she had never shown any interest in clothes, hair or make-up before he entered Number 10. As a barrister, she just had to put on a black suit every day and her hair was covered by a wig. She never painted her face. Now, suddenly, journalists were critiquing every public appearance she made.

It was incredibly frustrating, she confided. 'Because I was used to speaking and because I always had spoken, it was quite easy for me to carry on speaking about issues that didn't clash with government policy, and which were to some extent seen as safe issues because they were seen as female issues: the role of women, work–life balance, stuff about kids. But still you'd turn up and say something and all they'd want to talk about was what you were wearing, which is a bit insulting if you're an intelligent woman, because you don't expect to be judged by your clothes.'[13]

Is there any escape from this for women in public life? Julia Gillard and Ngozi Okonjo-Iweala interviewed eight women who had led countries (and in Christine Lagarde's case, the IMF and European Central Bank).[14] They all complained about the gratuitous focus on their appearance. Some resorted to the female equivalent of the uniform, wearing roughly the same thing every day so that it no longer merited attention. Think of Hillary Clinton's pant suits – or indeed Angela Merkel's.

There is a much easier solution, though, that addresses the problem at its root. If every political journalist simply asked themselves each time that they commented on a woman's looks, 'Would I have written this about a man?', the problem could be solved in five minutes. If we only applied the same standards to men and to women, this distracting and demeaning phenomenon would simply disappear. The media shape our perceptions so strongly that it would make a big difference to our attitude to women in power.

Women outside politics also have to worry about being judged on their appearance. Jess Brammar of HuffPost UK told me, 'I cut my hair shorter because I thought I couldn't be taken seriously if I had long hair. I remember when I was at *Newsnight*, for a while, I used to not get colourful manicures because I thought they made me look too frivolous. I wear colourful clothes. I wear dresses. And I sometimes wonder whether that has an impact on the way people perceive me.

'But actually, rather than trying to hide who I am, my femininity, the fact that I'm interested in fashion, I'd rather show that you can be super smart, a good journalist, run a team and also have those sorts of interests, because no male editor is ever maligned if he's interested in cricket or football. But until I worked with [the *Newsnight* presenter] Kirsty Wark, it had never occurred to me that you could be a senior, authoritative woman and talk about clothes and how much you love them. And she was really great about that. She just was like, "No, I love fashion. I find it really interesting and it's artistic." Yet it's completely dismissed, whereas you can be into other types of arts and creativity.'[15]

For younger women, this is even harder. Pandora Sykes, millennial author and cultural podcaster, told me, 'Being blonde and interested in colourful clothing becomes an easy shorthand for people to assume that what you do is frilly and frivolous and lacks depth. Certainly I've had people assume a lot of times that I wouldn't know about something or wouldn't be interested in something because of the way I look. Women who enjoy the outer façade aren't taken as seriously. The time you spend doing that is time that you could spend being cerebral.'[16] Yet if you don't spend the time making yourself look good, you are dismissed as a frump.

The podcast Sykes hosted until 2020 with Dolly Alderton, *The High Low*, was about high and low culture, with a lot of discussion of literary fiction. But people automatically assumed that it was about frivolous fashion, and it appeared in lists of the best fashion podcasts, even though the two presenters were adamant from the start that they wouldn't talk about clothes. 'Week in, week out,' said Alderton when it was still on air, 'we talk about politics, we talk about social issues, we talk about issues for minorities, we talk about feminist issues, and we are still being billed as a fashion podcast!'[17]

If a woman, particularly a younger one, is attractive, it is very hard for her to persuade people to listen to what she says rather than fixating on her looks. Laura Bates vividly remembers the picture editor of a national newspaper ringing her up to talk about a shot to accompany an article on her work. She takes up the story. 'He said, "The most important thing is to make you look as sexy as possible." And then said, "We have a spectrum of options, from a picture of you in a micro miniskirt, walking past a building site being whistled at, to you as a kind of sexy office vamp in a tight, sexy suit, with a stiletto heel perched on a man's throat." And it was fascinating, because it was in the context of an article about my work and about my expertise in this area. And he was very clearly saying, first of all, none of that really matters, we want you to look sexy – "the most important thing", those were the words he used – but also, interestingly, it revealed that his idea about a woman speaking about gender inequality was that she either had to be a victim or a man-hating bitch, and that there wasn't anything in between.'[18]

In some jobs, women can look incongruous, at least to start with, if these are positions that have always been associated with men. When Dame Sarah Mullally became Bishop of London, 132 men had held the post before her, over many centuries. Her predecessor, Richard Chartres – tall, with a pointed beard and a deep, sonorous voice – looked like an archbishop from the court of Henry VIII. Physically, he was a hard act to follow.

'I was very conscious that I just can't compete with that, that isn't who I am,' she told me. 'I have to have an integrity about who I am and I'm not going to change my voice very much. But cathedrals are big places, therefore I'm conscious there is a physicality about it, so how do I fill the space? I will stand up taller or put my shoulders back. I'm very conscious what I wear. How do you claim your presence without being bigger? Because you can't be that much bigger.'[19]

This was the five-foot-nothing Janet Yellen's problem too. It was apparently one of the main reasons why Donald Trump failed to reappoint her as Chair of the Federal Reserve, despite her being, according to the *Washington Post*, 'the most qualified Federal Reserve chair we've ever had and maybe the most successful Federal Reserve chair we've ever had.' I met her in Washington months later and she was still astonished by it. 'There was a story in the *Washington Post* that said the most important reason Trump didn't reappoint me is that I'm short and don't look the part of a central banker, to which I don't know what to say. I think he wasn't going to reappoint me anyhow, but it's just remarkable. To hear that really surprised me, but I guess it told me, yes, that stuff is still out there.'[20]

What does the research tell us about the effect of our looks on how other people rate us? Well, it definitely helps women to look 'put together'. The bad news for me is that women with curly hair are thought to be less professional than those with straight hair.[21] And the bad news for black women is that those with natural hair are rated as less competent and less professional than those with straightened hair. In fact, black women with natural hair were also the group least likely to be called for interview in this study. Black women often feel they have to straighten their hair for a job interview or keep it straight until they have been in the job for a while, so they can prove their competence before taking the risk of falling prey to this ridiculous hair bias.[22]

Grooming for women seems to be incredibly important. When investigating why attractive people of both genders tend to get paid and promoted more, University of Chicago professor Jaclyn Wong found that there are two aspects to what we consider 'attractive': what we're born with (our faces, our figures, our height) and what we can work on (hair, make-up, clothes, Botox).[23] Her research found that the latter makes the biggest difference to how women are perceived. Better grooming helps women succeed. As ever, though, they have to get the balance just right, because women with too heavy make-up are thought to be less competent, experienced and warm.[24]

They also have to be very careful not to dress provocatively, particularly if they are in senior jobs. Even undoing one more button on a shirt can make a difference, as can wearing a slightly shorter skirt.[25] Both of these make other people think that senior managers are less intelligent, trustworthy, responsible and authoritative. But it doesn't make a difference for more junior jobs, such as receptionist.

As women become more senior, their natural attractiveness may work against them. 'Once women get into managerial positions, positions of leadership, positions of power, beauty becomes a liability because our stereotypes around beauty are that they're incompatible with capability,' says Wong. 'So if you're too beautiful, maybe you're not that competent. Maybe you're a "dumb blonde". That's a lot more true for women than it is for men.'

If people find it hard to believe that a beautiful woman can also have a large brain, they feel uncomfortable when she proves that she does. Elizabeth Healey, an actor, told me the story of a movie casting audition. 'The director looked at my acting CV and then noticed my Ph.D. listed at the bottom under the "extras" section and he sort of laughed under his breath. He looked up at me, pointed at it and said, "This is a joke, right?" I said no, it was real, his face fell and I knew in that moment there was no way I was getting that role.'[26]

Beautiful women are also thought to be less trustworthy. Leah D. Sheppard of Washington State University and Stefanie K. Johnson of the University of Colorado Boulder mocked up articles about company layoffs that included photos of the managers announcing the job losses.[27] They then asked participants to rate the honesty of the leaders pictured, and decide whether they too should be fired. When the executive was a woman, people found her to be less truthful and more worthy of losing her job if she was also highly attractive. There was no such effect for men.

But if these are penalties for beauty, sometimes the bias can go the other way. There are some men who can't believe that an unattractive woman will be competent, or, at least, they know they don't want to share a boardroom table with her. A FTSE 100 chairman told me that he had recommended to his (male) CEO a highly talented but unfortunately rather plain businesswoman to be a non-executive director of the company. The CEO tried to veto her on the grounds that her credibility was undermined by her unattractive appearance. Thankfully, the chairman stuck to his guns and she turned out to be a great member of the board.[28]

We get such mixed messages here. Is it an advantage to be beautiful or a disadvantage? Confusingly for women, the bias seems to go in both directions. Dame Jayne-Anne Gadhia, former CEO of Virgin Money and now founder of the fintech company Snoop, believes that her extreme height – she is six foot two – has helped her, because it has taken sexual attraction out of the equation. 'I think it makes a difference if you are attractive or not and, broadly, because of my size, I'm not,' she told me.[29] 'If men are not sexually attracted to you, then it's easier to have an equal relationship. I think that that has made it more difficult for very attractive women to have the same level of gravitas.

'I remember when I was working at Royal Bank of Scotland and my boss at the time said to me, "Well, of course, one of the things that's good with you is that you have gravitas," and I went, "Do I?" Well, at my size, you can go into a room and be noticed, so yes, I suspect that makes a difference.'

And when she's talking to men, I put it to her, her eyes are level with theirs or she might even sometimes be looking down on them, whereas we average-height women always have to crane our necks upwards, and that, in itself, feels submissive.

'Yes, I'm sure that's true,' she acknowledged. 'I notice that occasionally, if I am talking to a very tall man, and I find myself looking up. And I do think, "Gosh, for a lot of my female friends, that's constantly how people are," and that's not my experience, so yes, that's different, for sure.'

Christine Lagarde, too, has said that she is too tall and too old to suffer sexism. 'It is hard to be sexist towards someone who is older and taller than you,' she says.[30] Being past the age of sexual interest from men may well be an advantage when it comes to women's authority, as long as it doesn't render them completely invisible.

Lesley Stahl is a veteran political journalist with CBS News's *60 Minutes*. She was the interviewer whose tough questions led Donald Trump to walk out before the end of his TV interview with her during the 2020 election campaign. We met in her New York apartment overlooking Central Park, and I asked her how long it took before she was taken as seriously as her male colleagues. 'I'm not sure I was ever treated with complete equality up until very, very recently, and I'm in my seventies!' she replied.[31] She believes that age is key: some men's resistance to female authority is bound up with sexual attraction, and that wears off with age. 'A headmistress of a girls' school told me she always had trouble with the fathers. If there was any issue with a kid, the father and mother would come to see her together and the father was in her face, poking his finger at her, telling her this and telling her that, "You can't do this, you can't do that." And she said, "Very recently, I realized I was having no trouble at all with the fathers. And I thought, wow, I really understand how to do this job, I must be really good at it, to the extent that these fathers aren't coming at me any more. I must be authoritative." Then she realized: "I'm over sixty-five, I can no longer have children, this biological issue between men and women is that they're not trying to dominate me because I'm past my reproductive years. I'm not a challenge, so all that junk goes away."'

Yes, age definitely helps. Nancy Pelosi, the eighty-year-old Speaker of the US House of Representatives, has talked to Stahl about this. 'She is very conscious of trying to show that a woman can be tough as nails, demanding, strong, steely, wear very high

heels and beautiful clothes, always look pretty and there's no con-
tradiction. She's really setting out to demonstrate that. She's said
that to me.'

It's very annoying that women should have to think so hard
about how they look. But the world – for now, at least – is as it is, so
if they want to be taken seriously, it's probably sensible for them to
be well groomed and not dress provocatively. Since it isn't clear
whether it's an advantage to be attractive or unattractive as a woman
at work, we can perhaps stop obsessing so much about the looks we
were born with. And we can console ourselves that, in some ways,
it gets easier with age.

Where we could achieve change very quickly and easily – if only
the will were there – is in the way the media portray women. Journalists should try to resist devoting so many more column inches to
women's appearance than men's. And editors should always 'flip'
the descriptors in articles and ask whether the same would be said
of a man. If not, just cut it out.

14

Shut your whore mouth!
The dangers of having an opinion and a vagina

'Men are afraid that women will laugh at them. Women are afraid that men will kill them.'

– Margaret Atwood

'The real problem is not denial, but **resentment** of female authority – a resentment which no woman should take as a compliment, since what is ultimately behind it is misogyny.'

– Deborah Cameron

I WAS AT A financial conference recently on International Women's Day, and was surprised and rather annoyed to see that the four-strong panel of economists was entirely male. It was the second all-male panel in a row, so I sent out what I thought was this pretty harmless tweet: 'Am at a conference on International Women's Day and there's a panel of four male economists and no woman. Really? There are so many great female economists.' To which I got the reply, 'Frankly, if my wife was anything like the Sieghart woman, I'd abuse her.'

It was the disproportionality of the response that brought me up short. Had I tweeted, 'Men should have their balls cut off,' I might have deserved such a reply. The mildest argument for greater equality these days ensures a (sometimes literally) violent backlash.

Before we get to the violent threats, though, let's just look at how women's authority is trolled on social media. When the Australian journalist Julia Baird tweeted about the media's different treatment of male and female politicians' private lives, she received this reply: 'And you have evidence of this or are you just being a bitter old sexist?'[1]

'Yes, I have written a Ph.D. on the subject,' she answered. 'So it's Dr Bitter Old Sexist, mate.'

There began a ridiculous Twitter spat. She was told by other Twitter users that she was an elitist snob, that five years of research were simply her 'opinion', that doctorates were no sign of intelligence and that she should be ashamed of herself.

She immediately changed her name on Twitter to Dr Julia Baird.

Trolling inevitably ensued. But so did replies from male academics who were baffled that anybody should notice or object. Dr Alan Nixon, a researcher in the sociology of religion at Western Sydney University, said, 'I've had Dr in my profile name since 2015 and have never been questioned on it.' Dr Stephen Maclean, an anatomy lecturer at the University of Edinburgh, wrote, 'I had no idea doctorate-shaming was even a thing!'

The British historian Dr Fern Riddell had a similar experience. Reacting to a newspaper deciding to restrict the 'Dr' title to medical doctors, she tweeted: 'My title is Dr Fern Riddell, not Ms or Miss Riddell. I have it because I am an expert, and my life and career consist of being that expert in as many different ways as possible. I worked hard to earn my authority, and I will not give it up to anyone.'

One man tweeted, 'If you have to tell people you're an authority or an expert then you probably aren't.' A 'David Green' suggested that her comments could 'legitimately be regarded as immodest'. Many other, much more offensive replies were posted, but Dr Riddell decided to start the hashtag #immodestwoman, which promptly spread around the world.

'You can clearly see that women have been taught to struggle with acknowledging their own authority and the huge backlash from lunatics online shows how women are taught to know their place,' she said.[2]

And sometimes this backlash escalates from the gentlest of provocations, such as Caroline Criado Perez getting rape and death threats for merely suggesting that we have a woman on one banknote to break the all-male monopoly.

Over one weekend, police gathered enough rape and death threats against her to fill 300 A4 pages. Most of these were not publicized in the media because they were too obscene; and, as a result, we didn't understand the true scale of the problem. So I think it's important to print some of these threats in full, because otherwise the sheer horror of them is lost. Here are just a handful of the thousands that she received. Look away now if you can't stomach it:

FIRST WE WILL MUTILATE YOUR GENITALS WITH SCISSORS, THEN SET YOUR HOUSE ON FIRE WHILE YOU BEG TO DIE TONIGHT.

I have a sniper rifle aimed directly at your head currently. Any last words you fugly piece of shit? Watch out bitch.

WOMEN THAT TALK TOO MUCH NEED TO GET RAPED

PUT BOTH YOUR HANDS ON MY COCK AND STROKE IT TILL I CUM ON YOUR EYEBALLS. DO AS I FUCKING SAY OR I'LL SLIT YA THROAT.

I'm going to pistol whip you over and over until you lose consciousness while your children watch and then burn your flesh

A BOMB HAS BEEN PLACED OUTSIDE YOUR HOME. IT WILL GO OFF AT EXACTLY 10:47PM ON A TIMER AND TRIGGER AND DESTROY EVERYTING

SHUT YOUR WHORE MOUTH . . .OR ILL SHUT IT FOR YOU AND CHOKE IT WITH MY DICK. OK?

No wonder two of her trolls were eventually sent to jail. Yet she was labelled a 'delicate flower' for complaining. This woman's life was in danger and she was understandably reduced to a nervous wreck. The onus is categorically not on women to develop a thicker skin; it is on trolls to change their vile behaviour.

On the day the attacks were at their worst, she says, she 'broke down completely', fearing that it would never end. 'By this point, it had been going on for a week . . . I struggled to eat, to sleep, to work. I lost about half a stone in a matter of days. I was exhausted and weighed down by carrying these vivid images, this tidal wave of hate around with me wherever I went . . . The psychological fall-out is still unravelling. I feel like I'm walking around like a timer about to explode; I'm functioning at just under boiling point – and it takes so little to make me cry – or to make me scream.'[3]

Women have come a long way in the past few decades. Society is far more equal than it used to be. Many men have embraced this, but a toxic few are fighting back vituperatively. They want women who speak to be silenced, put back in their place. It is dangerous to be in possession of both an opinion and a vagina these days, to lay claim to authority. If anything, the more authority women win, the more violent is the backlash.

We see this with sexual harassment. Women who are assertive or dominant or independent are the most likely to be harassed.[4] They are being punished for being 'uppity', for not knowing their place.

But the internet is something else. Women are twenty-seven times more likely to be abused online than men.[5] In 2015, the writer Alex Blank Millard changed her Twitter profile pic to a white man but kept the content – mainly woke tweets about sexism, racism and fat-shaming – the same. When she tweeted as Lady Alex, she had a pile-on of rape and death threats. When she commented on the very same things as White Dude Alex, she was retweeted, favourited, and even cited by Buzzfeed.

'For an entire week,' she wrote, 'I got to see what it is like to be treated with respect. As a man, I could use the same words and be met with discussion, with disagreement, or even nothing at all, in-stead of insults. I became an equal human being, one whose voice deserved to be heard.'[6]

Conversely, the fitness columnist James Fell never receives online threats, despite being a vocal feminist. 'You want yet another example of male privilege?' he writes. 'It's being able to voice your opinion and not face death and rape threats as a result.'[7]

There's a particular type of curdled misogyny that singles out women online. As the columnist Laurie Penny puts it, 'An opinion . . . is the short skirt of the internet. Having one and flaunting it is somehow asking an amorphous mass of almost-entirely male keyboard-bashers to tell you how they'd like to rape, kill and urinate on you.'[8]

In 2016, the *Guardian* analysed ten years of its comment threads, numbering 70 million, and found that of the ten regular writers who received the most abuse, eight were women (four white, four non-white) and two were men of colour.[9] Two of the women and one of the men were gay. The more traditionally male the section – for instance, sport and tech – the more women writers got abused. And who were the ten writers who received the least abuse? All men.

Amnesty International's Troll Patrol project used AI to survey millions of tweets received by British and American journalists and politicians throughout 2017.[10] Women of colour were 34 per cent more likely to be mentioned in abusive or problematic tweets than white women, and black women 84 per cent more likely.

Powerful women are particularly vulnerable. A recent study of American mayors found that 79 per cent had been the victim of harassment, threats or other psychological abuse.[11] Thirteen per cent were victims of physical violence. And one factor stood out above all others as a predictor of whether a mayor would be targeted: gender. Controlling for other factors, the researchers calculated that female mayors were more than twice as likely as their male counterparts to experience psychological abuse and almost three times as likely to experience physical violence.

In Britain, too, during the rows over Brexit, some female MPs were forced to move house and hire bodyguards. They were advised by police not to drive alone, not to travel after dark and not to run in the park.[12]

Occasionally, the perpetrators are senior, distinguished men in their own right. Who can forget the abuse that the young Democratic Congresswoman Alexandria Ocasio-Cortez received from her 65-year-old Republican colleague Ted Yoho in 2020? Here's how she described it in a searing speech in the House of Representatives: 'I was walking up the steps of the Capitol when Representative Yoho suddenly turned a corner, and he was accompanied by Representative Roger Williams, and accosted me on the steps right here in front of our nation's Capitol. I was minding my own business, walking up the steps, and Representative Yoho put his finger in my face, he called me disgusting, he called me crazy, he called me out of my mind, and he called me dangerous . . . In front of reporters, Representative Yoho called me, and I quote, "a fucking bitch".'

But you don't have to be a prominent woman in public life for it to happen to you. Karen Cohen, a former YouTube executive, says that even the most everyday videos by ordinary girls and women are subjected to vile abuse: 'Here were these creators putting up DIY fishtail braid videos, and there were people telling them, "I want to rape you" in the comment section.'[13]

Just being a woman with an internet connection can be enough. In 2006, researchers from the University of Maryland set up a bunch of fake online accounts and then dispatched them into chat rooms. Accounts with feminine usernames incurred an average of 100 sexually explicit or threatening messages a day. Masculine names received 3.7.[14]

And the women don't have to be saying anything provocative either. Women are being abused and threatened with sexual violence for commenting on bike-riding, comic book covers and soft pretzel recipes, says Emma A. Jane, author of *Misogyny Online: A Short (and Brutish) History.*[15]

It is a short history because this is such a recent phenomenon. As a journalist, Jane was used to receiving snail-mail letters taking

267

issue with the views expressed in her writing. But when she made the mistake of publishing her email address at the bottom of her column, suddenly the violent threats began piling in. 'I wondered what kind of person talked like this. What kind of person read through a newspaper and thought: "Hmmm. I don't appreciate Reporter X's writing. I think I'll send some hard-core porn-mail recommending a good, solid raping. Now there's a response that's both appropriate and proportionate!" Except that I wasn't Reporter X. I was *Girl* Reporter X. I began asking my male peers whether they were receiving emails from disgruntled female readers threatening pack sodomy, de-testiclization, and wall-to-wall sexual violence. And obviously they replied no.'

The men who send these violent, sexualized threats to women who dare to express an opinion are trying to impose a steep tax on entering the public sphere. The attackers don't engage with the arguments; they simply want to make it too costly for these women – and others who are deterred by the fear of being trolled – to join the national conversation. They want to silence women, so that the world hears only from men.

You can see this in the nature of the grotesque threats they make. There is an obsession with mouths and speech: cutting out tongues, fellatio, choking, oral rape, beheading, and all the things they would like to use to fill or shut a victim's mouth. Their fantasy is that women should be shut up – forcibly, if necessary.

And the dreadful truth is that it works. Women are advised not to speak out about the abuse they suffer, because that only 'feeds the trolls'. Women are terrified of going into politics or expressing views on TV, online or in print because they fear being trolled, doxxed or even murdered, like the MP Jo Cox.

Doxxing – publishing private details of a victim, such as her home address or phone number – can seriously endanger a woman's safety and wreck her life. Kathy Sierra, a software designer and game developer, was once one of the most high-profile women in tech. She received hundreds of rape and death threats on her blog. The trolls posted doctored photos of her being choked by underwear and with nooses next to her head. They circulated her home address and social security number and false statements about her being a former sex worker and battered wife. Her final blogpost read: 'I have cancelled all speaking engagements. I am afraid to leave my house. I will never feel the same. I will never *be* the same.' After that, she gave up her career and disappeared for years, not only from the online world but from offline public life as well. She later wrote: 'I had no desire then to find out what comes after doxxing, especially not with a family.'

Many women also have rape or death threats sent to their home address, with the message, either explicit or implicit, of 'I know where you live.' It is impossible to lead a normal life after that. Here's the experience of the *Guardian* columnist Van Badham: 'There was the man who told me he'd come along to a demo I was attending to slit my throat. Would you take that chance? The unknown number of people who followed me home from work and spied in my apartment from across the road, making observations on Twitter . . . And the other day, the packet of papers turned up in my house – my house! – with depictions of gang rapes and female genital mutilation, and the greatest of all threats unspoken: I know where you live. I've barely stayed the night there since. I'm moving soon.'[16]

Where does this hatred come from? What are the roots of this sulphurous misogyny? I must emphasize that it affects only a small minority of men, generally those who are deeply insecure about their masculinity, who have a sense of 'precarious manhood'. These are men who feel threatened by women, who feel emasculated by them, who feel that their masculinity is imperilled by women having power over them or even by encroaching on their domain. Some of them have trouble finding sexual partners, so-called 'incels', and blame women for rejecting them. Psychotherapy is very far from being an exact science, so each of the psychotherapists I talked to has his own theory, which isn't susceptible to proof. But here are some tentative answers – and I deliberately chose to talk to male psychotherapists, who were likely to have better insights into the male psyche.

Phillip Hodson, author of *Men: An investigation into the emotional male*, told me: 'The men most concerned with this issue are already to be found on the more insecure/worried/macho end of the classic masculinity spectrum. Emasculation literally means loss of the penis; more broadly, it implies castration. The problem concerns a woman possessing greater power than the man; or more power than he believes she ought to have; or more power in the wrong classes of activity than he thinks she should have.

'Instead of finding a powerful woman admirable (because she's a good role model), neutral (because bosses are supposed to give orders), or irritating (because she always seems peremptory), the man feels threatened. It very clearly involves a sexual theme and in some symbolic fashion and in his own mind he fears this state of affairs will result in the theft of his potency. The bottom line is that a woman who possesses greater power and authority than this type of man tends to fill him with a sexual fury. Not only could he never reasonably expect to become her lover, but even if she did condescend in that direction, he probably couldn't "perform" with her.'[17]

When men's masculinity is threatened, they can easily turn aggressive. Psychologists Jennifer K. Bosson and Joseph A. Vandello from the University of South Florida did an experiment in which they asked some men to braid hair and others to braid rope.[18] Afterwards the men were given the choice of either punching a bag or doing a puzzle; the hair-braiders overwhelmingly chose the former. When one group of men braided hair and others didn't, and they all punched the bag, the hair-braiders punched harder. When they all braided hair and only some got to punch, the non-punchers showed more anxiety on a test. Aggression, write the authors, is a 'manhood-restoring tactic'.

The difficulty for men is that manhood is seen by men as something to be earned rather than simply bestowed by growing up. Girls automatically become women; boys don't automatically become men. In some cultures, indeed, they still have to pass painful initiation tests to gain their status as a man. Metaphorically, they still do. Think of Rudyard Kipling's poem 'If': after a whole string of strictures, which include being calm, brave, patient, confident, resilient and determined, he declares that only if you pass them, 'You'll be a Man, my son!'

Not only does manhood have to be earned, it can also easily be lost by acting in a feminine (hence inferior) way. It is precarious. Womanhood, by contrast, can't be lost by behaviour alone. And men are the main enforcers of gender role violations, so they are forever worrying what other men will think of them. Will they think I'm gay? Sissy? Effeminate? A big girl's blouse?

What creates this insecure masculinity in some men? According to psychotherapists, it goes back to very early childhood. A baby (male or female) is utterly dependent on its care-giver, usually its mother, not just for food and comfort but for its very life. And, as Adam Phillips puts it: 'Everybody loves and hates the people they most depend on; traditionally, it was a woman, though it isn't always so now. What they really hate, psychoanalysis suggests, and sometimes insists, is the part of themselves that is dependent. And they protect this part of themselves by attacking its object: men don't hate women – they hate their need for women.'[19]

But surely baby girls are just as dependent on their mother as boys? Yes, but as girls grow up, they don't have to split away from their mother, to define themselves as something different, in the way that boys do. Adam Jukes has written many books on this

271

subject, including *Why Men Hate Women* and *Is There a Cure for Masculinity?*. He explained the difference to me. 'The little girl, in order to become a woman, just has to copy her mother. The little boy wants to be close to his mother, doesn't want to let her go. In order to be a boy and then later a man, he has to identify with his father, but that's not a natural process because he started off identifying with his mother. Little boys not only have to start identifying with the father; they have to dis-identify with the mother.

'There's a schism in the male personality because once a boy makes that decision to dis-identify with his mother, you can't get rid of the identification, all you can do is repress it, all you can do is deny it and push it down, but it's always there, the yearning, the dependence, the unrequited love, it's all still there.

'The further along the masculine spectrum a man is, the more afraid he's going to be of women, and therefore the more important it's going to be for him to be in control of her because the one thing she can do that terrifies him the most is to take him through that gap from the masculinity into the repressed yearning, the failed dependency or the vulnerability, the dissolution.'[20]

How often have you heard a man complain of a female boss, 'She reminds me of my mother'? It's never a compliment. Yet we rarely hear a woman say, 'He reminds me of my father.' And if she did, we wouldn't assume it was disparaging.

For many little boys, the difficulty is compounded by the absence or distance of their father. As Nick Duffell, founder of the Centre for Gender Psychology, put it to me: 'The father is very often at a distance, so you've got this built into the male psyche, I believe, from very early on.'[21] This distance makes it harder for these sons to feel comfortable in their masculinity because they have never been close enough to their father to model themselves upon him. And it allows them not just to dissociate from their own feelings, but also from the feelings of the women they encounter – or bully.

Another factor, according to Phillip Hodson, is that some boys and men suffer from what we might call, in a nod to Sigmund Freud, womb envy. 'A boy,' he told me, 'will realize that he – compared to his sister – has a biological deficit. Whereas girls can become both scientists and Presidents of the Royal Society, they can additionally – or alternatively – choose to give birth to their replacements in a potentially fulfilling act of physical creation. By contrast, boys are incapable of this feat, they are excluded from the mystery, and in basic evolutionary terms, their seeds are less valuable than their sisters' eggs.

'As a consequence, boys – agitated by testosterone – are driven by default to seek some form of power in the world to fulfil their existence and justify their effectiveness. How else to hand your name on and prove that you've been here? Even if your economic circumstances are oppressive, or your family is criminal, you will try to rise in your gang of mates, earn their respect and acquire some form of reputation. Better to have had bad attention than none. Better to be "El Chapo" for five minutes than some nameless peon.

'I would suggest it is not girls who suffer from penis envy but boys. It is boys who perpetually measure their performance in or out of bed; males who become obsessed about the pecking order and whether they are entitled to an executive parking space. Men who count and keep score. And males who consciously or not envy and resent women's biological creativity. Forced to compete in the outside world, terrified of rejection, lacking any of the attributes of James Bond, Superman or Simon Cowell, the young male does not even have the option of falling pregnant.

'So men turn to this stressful quest for power over the planet, their brethren and themselves. And in the process, they tend to belittle the efforts of women to claim to have any say in the same arena.'[22]

These insecure men usually suffer from an acute case of male entitlement: the notion that men somehow *deserve* to be in a position of superiority over women and that women have a *duty* to serve them, stroke their egos, and be submissive. When women refuse to play the submissive role, to conspire in this notion of inferiority, these men lash out.

'Such men,' says Hodson, 'either have struggled to get mothers and sisters to behave like this – so now impose such demands as adults – or enjoyed much cosseting from mothers and sister figures and so regard these assumptions as normal. In both cases, I suggest they have been deprived of *ideal* nurturing (which creates a more secure confidence). So instead of being able to stand on their own two feet and sustain criticism by drawing a boundary around their own personal identity, they define themselves as a broader archetype of a general social masculinity to be supported and defended at all costs.'

The sociologist Michael Kimmel calls this 'aggrieved entitlement'.[23] Now that women have entered domains in which men in the past were in competition only with other men, there is (from some men) a huge amount of resentment. As Kimmel says, on the basis of the many interviews he conducted with white men, when a black woman is hired over a similarly qualified white man, the man is prone to complain that the woman took his job. Why *his* job, not *that* job? Because the man feels unjust and outdated patriarchal (and, in this case, racist) entitlement.

Yet it's not as if women want to take over the world and push men into a position of inferiority. This may be what a lot of men fear, but it's not how the female psyche generally works. Most women just want to be given a genuinely equal chance, alongside men. They don't want to turn the tables and dominate men, to make them suffer.

But alt-right movements are encouraging insecure men – and teenage boys – to believe that this is what women want and to conclude that feminists have caused everything that is wrong with their life. They can't get a job? Blame uppity women. They can't get a girlfriend? Blame man-hating women.

Laura Bates went deep undercover in the online so-called 'mano-sphere', where the worst misogynist messages are propagated, to research her book *Men Who Hate Women*. She is horrified by the extent to which teenage boys are being seduced by this message. 'One of the most disturbing things that I found,' she told me, 'was that these groups are actively recruiting and indoctrinating boys as young as eleven, and in very clever and effective ways. They are using everything from viral YouTube videos to Instagram memes, to slideshows that they disseminate amongst their members, to body-building websites, to gaming livestreams, to private chatrooms, to send out the message to teenage boys that there is a feminist con-spiracy at the heart of our government that is actively discriminat-ing against white men. They claim that white men are in danger of extinction; that women are taking men's jobs and their livelihoods; that some 90 per cent of rape allegations are false and that men are at enormous risk from them; that tens of thousands of men are raising children that aren't their own because unfaithful women are cuckolding them and then forcing them to financially provide for somebody else's child; that the gender pay gap is a myth. I mean, the list goes on. And the impact that they're having is really quite shocking, but it isn't generally known.

'So we have this situation where people are very aware of other forms of grooming and radicalization, where we have a national strat-egy to try and prevent that, where teachers are trained and required to be on the alert to report any hint that somebody has come into con-tact, for example, with Islamic extremism online, but where the vast majority of teachers and parents have no idea that this other form of extremism even exists. And exactly what is happening to these boys is radicalization and grooming. But we don't use those words to describe it when the form of extremism in question is extreme misogyny.'[24]

This radicalization of teenage boys, which started only in the late 2010s, is having a dreadful effect on teenage girls too. Bates has been giving talks in British schools twice a week for nearly a decade and has really noticed the change. 'In a school where these ideas are particularly widespread or ingrained, the most arresting feature of the teenage girls at that school is their silence. They won't put their hand up, they won't ask questions, they won't engage with the conversation at all, because they're aware that to do so would mean being branded as a feminazi, as a man-hating bitch. For example, in one school, a girl who did talk about these issues and made some very reasonable and gentle points about gender equality later sent me copies of text messages she received from the boys in her year afterwards, accusing her of being a bitter man-hating feminist, accusing her of being a slut who regretted her own bad decisions, and saying things like, "It's not our fault, we're just biologically superior to you, you just need to learn that that's the way of the world." And I think that's quite shocking, these really outdated beliefs coming out of the mouths of teenage boys in 2020. But that's the kind of thing that we're dealing with. And it's breeding in teenage boys a dehumanization of their female peers, but also an intolerance to any other kind of viewpoints, to open-mindedness, to debate, which makes it really difficult then to reverse the problem.'

And this message is, terrifyingly for women, playing out in populist politics too. Look at the slogans on Trump merchandise when he was running against a woman in 2016: badges saying 'Trump that bitch' and 'Life's a bitch: don't vote for one', a badge depicting a boy urinating on the word 'Hillary', a T-shirt showing Trump as a boxer having just knocked Clinton to the floor of the ring, where she lies face-up in a clingy tank top. Can you imagine any of these images being used of Joe Biden? The worst he gets called is 'sleepy' or 'dumb'.

Mind-bogglingly, 49 per cent of Trump voters in 2016 said that men suffered 'a great deal' or 'a fair amount' of discrimination, compared with only 30 per cent who said the same of women.[25] (They also claimed that men had it harder than gay people, immigrants and African-Americans.) It's hard to get your head round this cognitive dissonance. You only have to look at the world to see that it is still largely run by straight white men. You can hardly deem it 'fake news' that every American president has been a man or that 93 per cent of Fortune 500 CEOs are male.

Yet these hostile sexist attitudes are what right-wing populist nationalist leaders feed off. Brazil's President Jair Bolsonaro told a congresswoman, 'I would not rape you, because you are not worthy of it.' He has called the secretary of women's policy 'a big dyke' and says women shouldn't be paid the same salary as men. When he voted for the impeachment of Brazil's first female president, Dilma Rousseff, who had been tortured under the military dictatorship, he dedicated his vote to one of her torturers. Meanwhile, crowds at Bolsonaro rallies chant that they would feed dog food to feminists.

In the Philippines, under President Rodrigo Duterte, the attacks on women are just as vicious. He 'joked' that soldiers enforcing martial law on Mindanao Island could rape up to three women with impunity. When an Australian missionary was gang-raped and killed, he said she was 'beautiful', so he should have been first in line. He told soldiers to shoot female rebels in the vagina because 'they are nothing without it'. In Italy, Matteo Salvini, the leader of the Lega Nord Party and former deputy prime minister, compared the female president of the lower house of parliament to an inflated sex doll.

Populist leaders want to put women back in their box, by restricting access to birth control and abortion and by encouraging women to be wives and mothers, ideally of lots of children. Viktor Orbán, Prime Minister of Hungary, is frequently dismissive and insulting about women. When asked why he had no women in his Cabinet, he said that few of them could cope with the stress of politics. But he has promised that women who have four or more children will never have to pay income tax again. Meanwhile, Poland's nationalist government has run advertisements urging Poles to 'breed like rabbits', has banned the over-the-counter morning-after pill and has almost completely outlawed abortion.

These messages are popular with a lot of men, and with some socially conservative – and often Catholic – women. But they pose a huge threat to women's reproductive rights and threaten to reverse the progress that women have made in the workplace over the past few decades. What it shows, depressingly, is that women can never afford to be complacent. Progress isn't automatically irreversible and women have to keep fighting, sometimes just to stand still.

This backlash against women, their voices and their authority, is deeply depressing and in some cases very scary. But no struggle in human history has been won without pushback, sometimes violent. For some men, women demanding a voice seems to overturn the natural order of things. We can change that natural order only by having more women in senior positions so that it seems more normal; and in future generations, by fathers spending more time with their families and parents sharing authority equally, so that boys can get closer to their fathers and children don't grow up believing that men have a right to boss women around.

We also need to demand that social media companies take trolling more seriously. And schools, parents and governments must urgently start to take action against the grooming and radicalization of teenage boys in extreme misogyny.

It will be a slow and painful process, but giving in to the toxic misogynists is surely the worse and more cowardly choice.

15

No need to despair
We can narrow the authority gap in one generation

'YOU WON'T EVEN KNOW YOU'RE DOING IT'

'The greatest barrier to gender equality is the fact that we have a male-dominated world, and many men don't even realize this. Many men take for granted the world as it is, and they look at gender equality as almost an aberration.'

– Antonio Guterres

S O WHAT HAVE we established? The authority gap is real and it damages women's standing in the world, and their pay and promotion prospects. We often don't notice that we are contributing to it, due to the tricks that our bias plays on our brains. Because boys and men are brought up to be more confident and to take up more conversational space than girls and women, we incorrectly judge them to be more worthy of respect and authority. Yet women can't compensate for this discrepancy without being punished in the process. And the gap is widest of all for women who are also disadvantaged by their race, class, sexuality or disability.

Often, men don't read or listen to women's voices in the first place, which makes it even harder for women to gain authority. And the gap is perpetuated by the way women and men are portrayed in the media, advertising and broader culture. This has, however, started to improve, as more female experts are quoted in news stories and female directors and scriptwriters create more authoritative parts for female actors to play.

We are still quite resistant to female leaders gaining power, though this is gradually lessening. And there is a fierce backlash from a small minority of men to women having authority at all.

But I don't want to leave readers feeling gloomy, because there are two pieces of really good news:

1. There is *a lot* that we can do to narrow the authority gap, and
2. As we've seen, the world will be a better place, *for men as well as women*, if we do.

In this chapter, I'm going to enumerate the ways in which we all can do something – as individuals, as partners, as parents, as colleagues, as employers, as teachers, as journalists, as governments, and as society – to make the gap smaller and, eventually perhaps, get rid of it altogether.

But nothing will change unless we acknowledge that the gap exists and that we want to do something about it. We are always far readier to spot bias in others than in ourselves.[1] For men, in particular, it can be challenging to admit that there is a problem. Male scientists, for instance, who have been trained all their lives to analyse and interpret evidence objectively, are still prone to evaluate research on gender bias less favourably than women are. This is particularly true among men teaching STEM subjects at university.[2] And the research itself is funded less often and published in less prestigious journals than research on race bias.[3] Gender bias still seems to be an inconvenient truth for some men.

These men are probably suffering from what Tony Hockley of the University of Oregon calls *solution aversion*.[4] As he explains it, 'The concept of solution aversion is the idea that people are motivated to deny problems and the scientific evidence supporting the existence of the problems when they are averse to the solutions.' If the solution to the authority gap were simply that men had to cede power to women, I can see why they might be averse to losing their privileged position in the world. I don't deny that *some* rebalancing of power is called for, but there are so many other ways of narrowing the gap that don't threaten men, and others that positively benefit them. So please read on.

I am going to suggest innumerable ways in which we can help to close the gap, and it will be impossible to remember to do all of them, all the time. It might be worth choosing a few to start with and revisiting these pages over the next few months or years to add to their number and remind yourself of what you committed to do. You may be surprised and pleased to find that the early ones have by then become automatic.

What can we do as individuals?

Our everyday interactions with women provide incessant examples of the authority gap in action. Every time that we ignore what a woman says, look over her shoulder for someone more interesting to talk to, or talk animatedly to her partner and refuse to include her in the conversation, we are explicitly demonstrating our implicit assumption that women are less interesting than men and that their views are less important and valuable. We may do this instinctively or reflexively, but we *can* change our behaviour, just as we can stop biting our nails or slumping at our desk. It just requires awareness of the problem and a bit of effort and practice.

I know how hard this can be. I am white and therefore privileged by my race. It's easy for me not to notice that I never have to worry about being stopped by the police for simply driving my old Mercedes. When I read the African-American Claudia Rankine describing how she walked into her daughter's parents' evening, saw a mass of white teachers and wondered whether any of the white parents had noticed the lack of diversity, I realized that I was one of those white parents.[5] I would have spotted if all my daughters' teachers had been men, but I didn't clock their overwhelming whiteness.

But my eyes have been opened by reading her, and many others', writing on race. I don't always get it right, but at least I am trying to unlearn some of my bad habits and default assumptions. I hope that readers of this book will have been similarly jolted into recognition. Because there is so much that we can do to narrow the authority gap.

So what can we do as individuals?

- We can accept that, however liberal and intelligent we are, we probably suffer from unconscious bias, whether against women, or people of colour, or people of a different class, or a different country, or a different sexuality. If we're extroverts, we may be biased against introverts. If we live in the country, we may be biased against city-dwellers, or vice versa.

- We can't stop this unconscious bias or put a lid on it. We don't need to feel ashamed of it. But we can recognize that the bias is based on incorrect assumptions and then correct for it. The more we become aware of our bias in our everyday interactions, the easier it is.

- We should resist starting, when we first meet her, with the default assumption that a woman will be less knowledgeable, competent or interesting than a man. When we're assessing a woman's ability, we can ask ourselves if we'd think the same if she were a man.

- We can try to integrate the evidence that women are just as intelligent, qualified and authoritative into our working assumptions about people.

- We can notice if, when walking up to a man and woman together, we address the man first.

- We can listen as actively and attentively to women as we do to men.

- We can notice if we interrupt women more than men and, if so, try not to.

- We can notice if we are challenging women more than men and, if so, ask ourselves why.

- We shouldn't assume that a woman will know less than a man about a stereotypically male subject. Just because fewer women than men, on average, follow sport doesn't mean that any individual woman won't be an expert on it.

- If we're male, we can try to internalize the notion that there is nothing humiliating about accepting a woman's expertise or authority and allowing her views to influence ours.

- If we're a woman trying to gain authority, we can use humour and warmth to deflect any hostility.
- We can notice how much of a conversation we're taking up. If we're hogging the conversation and talking much more than a woman, we can row back, ask her some questions and allow her half the conversational time.
- When talking to a woman of colour, a disabled woman or a woman of a different background or sexuality, we can be extra-vigilant about our biases.
- If we find a competent woman unlikeable, we can ask ourselves why. Is it her problem or ours? Would we feel the same way if she were a man?
- If we find ourselves judging people by their accents, we can try to listen to the content of what they're saying before we make up our mind about them.
- We can check the adjectives that come to our mind when we describe a woman. Would we use them for a man?
- If we are male, we can actively seek out books, films and TV programmes about or by women. We may be surprised how much we enjoy them.
- We can put pressure on media organizations, advertising bodies and the film industry to treat women more equally.
- We can look at who we follow and engage with on social media and try to redress the balance if it is many more men than women.
- We can stop mistaking confidence for competence. We shouldn't assume that the man with the loudest voice in the room knows what he is talking about. We shouldn't automatically believe a woman when she is being self-effacing.
- If we are female and underconfident, we can challenge ourselves to speak up and take on harder tasks.

- We can draw out a woman who is talking less than the men around her.
- Men can call out other men when they are being sexist and tell them it isn't acceptable; in fact, it is as obnoxious as racism. You may find you have more male allies than you think when you do this: research suggests that men over-estimate the sexism of other men.[6]
- Women can steel themselves to ask questions and make contributions even if they're not entirely sure of their ground.
- We can try not to judge women so much by their appearance or the pitch of their voice. When we find ourselves doing it, we can deliberately set that judgement aside.
- Most importantly, we can treat every single person we meet as an individual, and not judge them through a warped template of dated stereotypes.

What can we do as partners?

Women bear so much more of the unpaid workload than men – about 60 per cent more in the UK – that it is far harder for them to advance at work at the same pace. They often can't work the extreme hours that their male colleagues do if they have caring responsibilities at home. They may not be able to relocate for work unless they have an understanding partner. If they work part-time, they are usually overlooked for promotion. And, as well as carrying all their work responsibilities in their head, they have an equally demanding list of to-dos relating to every member of their family and everything that needs taking care of at home.

If these chores aren't shared equally between partners, the load on the woman becomes intolerable. But too often men in straight relationships expect their partners to shoulder this burden, even when the women themselves are in high-powered careers. A survey of Harvard Business School graduates found that, although the women started off expecting their careers to be as important as those of their partners, men were much less likely to share that view – even among millennials.[7] Until we have equality at home, we can't hope to have it at work.

Malala Yousafzai's father, Ziauddin, has put this into practice: 'At home, my wife and I treat our three children as equals and we try to demonstrate a more balanced partnership. Our children see me cook, clean the house, and pick them up from school – tasks too often seen as women's work. I was a feminist before I even knew the word. We are not perfect, but I hope one day our children will take the best of what we've taught them and work to make it even better . . . In one generation we transformed our family from a patriarchal one to an egalitarian one.' No wonder his daughter has grown up to be so brave and special.

So what can we do as partners?

- We can start from the premise that this is a relationship of equals.
- If we are in a relationship with a woman, we can give equal weight to our partner's career, if that is what she wants.
- We can always treat what she says with respect.
- We can take equal responsibility for the unpaid work at home and not see it as 'helping' her. This includes taking on an equal share of the planning that is involved in running a household and a family.
- We can be prepared to work flexibly, if necessary, once children are born. Men who work flexibly are more satisfied with their work–life balance than those who don't.[8]
- Men can take as much parental leave as possible. This not only helps the mother to continue working but has huge benefits to the father and the children too.

- If we are a woman in a straight relationship, and we want our partner to contribute equally at home, we can resist the temptation to criticize the way he does chores or looks after the children.
- If we are female, we can take Sheryl Sandberg's advice: 'Make your partner a real partner'. Choose him or her with care, and before you commit to each other, have those important conversations about how you will share responsibilities at home.

What can we do as parents?

It is going to be very hard to abolish the authority gap in this generation, as we are all fighting the prejudices that were instilled in us as children. But it should be possible to bring up a new generation that suffers much less from that handicap. Of course, we can't control what our children experience at school or what they absorb from their peer group or wider society, but we can do our best to equip them to challenge stereotypes from outside. And we can understand that the way we bring up our children – and particularly our sons – will probably have the strongest influence on how they will treat girls and women as they grow up. Almost all the powerful women I interviewed said that they were treated equally to their brothers, that their parents had high expectations of them, and – crucially – that their father really believed in them.

So what can we do as parents?

- We can make sure that our children see us, as parents, having equal authority at home.
- We can share the household chores and childcare equally. Research shows that fathers who do this bring up daughters who are more ambitious about their future careers.[9]

- We can treat our sons and daughters as absolute equals.
- We can raise our sons to respect girls and women.
- We can try to counteract our sons' beliefs that boys are superior. When the authors of *Still Failing at Fairness* asked American middle-school boys what was the best thing about being a boy, the second most popular answer (just after sport) was 'not being a girl'.[10] It is this sense of superiority that allows the authority gap to persist.
- We can instil confidence and self-belief into our daughters.
- We can avoid perpetuating stereotypes such as telling our daughters that they are pretty and our sons that they are clever.
- If we think our son is more intelligent than our daughter, we should check if this is objectively the case before believing it to be true. We may well be biased.
- We can offer toys, games and pursuits of all kinds to both sons and daughters: we can teach our sons to cook and our daughters to mend cars, as well as the other way round.
- We can encourage brave physical activity, such as climbing trees, for girls as well as boys.
- We can avoid bias creeping into the chores we expect them to do.
- We can encourage boys to read books about girls.
- If we can't find enough books with interesting female characters, we can change the gender as we read aloud to small children.
- We can point out and discuss sexist tropes and assumptions in the TV and films we watch with them.
- We can imbue them with the belief that sexism is as unacceptable as racism and other forms of discrimination, and we can instil in our sons the moral courage to challenge sexism in their peers.

- We can teach our sons to respect the views of girls, call them out if they interrupt their sisters, and encourage our daughters to speak up, inside and outside the home.
- We can encourage our children to consider a wide range of subjects at school and university, and non-stereotypical careers afterwards. Why shouldn't your son be a teacher or your daughter an engineer?

What can we do as colleagues?

For women, encountering the authority gap in their everyday social lives is irritating; encountering it at work is infuriating, because they know how damaging it is to their prospects of advancement. It is at work that the authority gap is at its most pernicious. Part of the responsibility for narrowing it lies with the organizations for which we work, and I'll come on to that next. But we can each individually, as colleagues, do a huge amount to narrow the gap for the women with whom we work.

So what can we do as colleagues?

- We can affirm what female colleagues say at meetings. If they make up only 20 per cent or 40 per cent of a group, they are less than half as likely as men to win approval from the other members and much more likely to be interrupted.[11] No wonder they feel less inclined to offer an opinion.
- If we are chairing a meeting, we can make sure that women are given equal talking time, and if they make a good point, we can say so. We can call out men for either talking too long or interrupting women. If a woman makes a point at a meeting and is ignored, and a man

makes the same point later, the chair can remind the meeting that the woman thought of it first.

- We can note that if a woman chairs a meeting, other women will contribute much more.[12] If decisions have to be reached by unanimity, so that everybody's opinion matters equally, women will also speak up more.[13]
- If we are chairing a Q&A, we can call on a woman to ask the first question. Research suggests that if we do so, more women will be inclined to follow suit.
- If we are a man, we can socialize with female as well as male colleagues, ideally at lunchtime if they are a parent, as they may have to rush home after work.
- If we are doing a performance evaluation, we can be aware that evaluations of women tend to be shorter, less positive, vaguer and dwell more on their personality than their achievements. One way of eliminating bias for evaluations and after job interviews is to use 'structured free recall': spending five minutes writing down the candidates' positive traits and then five minutes writing down the negative ones.[14] We then compare these with what we are looking for in a candidate for a job or promotion.
- We can reward hard work, preparation and attention to detail more than blagging.
- We can be careful of the adjectives that come to mind when appraising a female colleague. Do we think she is 'abrasive', 'strident', 'bossy', 'shrill', 'aggressive' or 'unlikeable'? If so, we have learned by now that this may tell us more about our status incongruity discomfort than about the woman herself.
- Before interviewing a woman for a job, we can spend a few moments visualizing in detail a woman in a leadership position. This helps to mitigate our bias. In job interviews,

we can be careful of the questions we ask. Women are taught not to self-promote and know that people may dislike them if they do. So asking a question such as, 'Tell me about a personal or professional accomplishment that best shows your strengths' can be problematic. A woman might be wary of boasting and tell us that she is proud of her children, while a man will give a work-related answer and seem more impressive.

- We can reduce bias by evaluating several people at the same time against each other and against a specific set of criteria.[15]
- We can undergo unconscious bias training. It won't necessarily change our behaviour unless we commit to action afterwards, just as a session on the evils of sugar and fatty foods won't on its own make us lose weight. But it does help to raise our awareness.
- We can check that we are giving challenging assignments to female as well as male colleagues. In a business, we can make sure that women get operational roles with profit-and-loss responsibility, as well as ones in marketing and HR.
- If we find ourselves thinking that a woman is talking 'too much', we can try surreptitiously timing her to see whether it's perception or reality.
- We can resist the assumption that a woman in her twenties or thirties will go off and have children and the temptation to penalize her accordingly. Men of that age are just as likely to leave their jobs. When a woman does get pregnant, we shouldn't automatically assume that she will be less committed to her career or will want softer options at work. We should ask her first.
- We should allow employees of both genders to work flexibly if they want to and not punish them in career terms if they do. The coronavirus pandemic has shown us

that workers – if their jobs allow it – can be just as productive, if not more so, working at home or during the hours that suit them.

- We shouldn't penalize female colleagues for being over- or under-confident. We need to understand that they are walking a tightrope that doesn't apply to men.

- We shouldn't assign all the 'feminine' office tasks to women – making the coffee, organizing the Christmas party – or hold it against them if they don't want to do them. Be aware that men are rated more highly for helping colleagues, and women are rated more negatively for not helping.[16]

- We can make sure we give helpful and specific feedback to female as well as male colleagues. In a study of 200 performance reviews in a tech company, researchers found that women were much more likely to receive vague praise than were men, including unhelpful comments like 'You had a great year.'[17] Men were more likely to receive developmental feedback, linked specifically to business outcomes. When women did receive developmental feedback, it tended to relate to their personalities rather than to their performance.

- We can be aware that men tend to reward other men more highly at work and, if we are male, we can try to correct for this. A study comparing objective algorithms with human evaluations found that 70 per cent of men rate men more highly for achieving the same goals as women, rising to 75 per cent for men in senior positions.[18] The algorithm rated men and women equally for similar performance, as did other women.

- We can resist putting women in teams in which they are the only woman or they are vastly outnumbered by men.

They will perform worse, contribute less, speak out less and feel more anxious. The men will also behave worse towards them.[19]

- We can try not to be harsher on women than men when they make mistakes. Research suggests that this often happens, particularly in traditionally male occupations.[20]
- We can actively encourage women to seek out promotions. Sometimes they are eminently qualified but don't have enough self-belief.
- We can think very carefully about what characteristics we expect from men and women when we are promoting or hiring. Often we look only for competence from men but a whole range of other characteristics as well as competence for women.[21] We can set out the criteria we want the successful candidate to meet in advance and judge all applicants against the same criteria. In one study, people evaluating candidates for the position of police chief were asked whether education or experience was more important for the job.[22] When the male candidate had more education, they said education was more important. When the female candidate had more education, they preferred experience.
- When we write letters of recommendation, we can make sure that we use the same kind of adjectives, and comment on the same matters, for women as we do for men. Letters about women are more likely to raise doubts, to contain faint praise and to use 'grindstone' words such as 'hard-working', 'conscientious', 'meticulous' and 'diligent'. Men's letters are more likely to have words such as 'successful', 'accomplishment' and 'achievement' and more uses of adjectives such as 'excellent', 'superb', 'outstanding' and 'exceptional'.[23]

- If we are men, we can try to re-examine what seems unfair to us. Katherine W. Phillips, a professor of organizational management at Columbia University, tells the story of meeting a group of new managing directors at a bank in New York. 'One question they asked was, "How do you explain to the white man with equivalent qualifications to a woman or a person of colour the decision to hire or promote them instead of him?" And I said: "Well, what do you say to the woman or person of colour who was equally capable? Why do you assume that the position belonged to the white man?" '[24] Entitlement can be as unconscious as bias.
- Men can refuse to appear on all-male panels at conferences. In general, men can see themselves as allies of their female colleagues and take part in diversity and inclusion initiatives so that women don't have to do all the heavy lifting.
- If we are male, we can ask for a female 'reverse mentor', who can help us understand how what we're saying or doing, probably inadvertently, can contribute to the authority gap.
- We can be aware of the dangers of affinity bias: preferring people like us. So if we are male, we can make a point of bringing on a promising junior female colleague, rather than instinctively picking a young man. The male-to-male advantage in promotions (a male manager promoting a male junior) accounts for a third of the gender pay gap.[25]
- Men can call out sexism when they see it at work. Not only does this make the sexist think twice before doing it again; it makes the female victim feel better about herself too.[26]
- Men can notice if they have reflexive hostility towards a female boss and try to correct for it. There is nothing

humiliating about having a woman in authority over you, and you may find that she is a better, more inclusive boss than a man.

What can we do as employers?

We used to think that there was a 'glass ceiling' at work that was impossible for women to crack. This is an unhelpful metaphor. For a start, some women do make it all the way to the top. But, more importantly, what prevents other women from succeeding is not just one obstacle but an accumulation of small disadvantages all through their career, many of which are due to unconscious (and sometimes conscious) bias.[27] From the first promotion to manager, which is so much quicker for men than for women, to all the other, myriad ways in which women are held back unfairly, the compounding effects of these setbacks multiply into a big difference over the course of a working life. This means that there is no one solution to narrowing the authority gap at work: what is needed instead is an accumulation of small solutions. Together, though, they can make a big difference.

So what can we do as employers?

- We can keep meticulous track of how women are doing in our organization compared with men. We should disaggregate the data to include intersectional factors such as race, class, sexuality and disability. Only if we know the facts can we act on them.
- We can make sure that managers all the way through the organization are aware of the disadvantages that women face and are held to account for the success they have in addressing them.

- We can actively encourage flexible working, at a senior level too. We must make sure that employees – male or female – are not punished for taking up this option, and use senior men who work flexibly as role models for more junior ones. We can advertise all jobs as flexible too.

- We can pay particular attention to employees, male and female, at the hinge point in their lives when they become parents. Don't assume that they will want to soft-pedal their careers but, if they do, make sure you are ready to promote them again when they want to re-engage. Be ready to believe that they are still just as committed to their job.

- We can take gendered words out of our job advertisements. Don't ask for 'competitive', 'assertive' or 'ambitious' applicants, let alone 'a ninja coder who wrestles problems to the ground' (true example).

- We can state in the job advertisement that salary is negotiable. It encourages women to negotiate.[28] (But then we mustn't allow our bias to find that woman unlikeable if she does.)

- We can use 'blind' CVs and application letters so that hiring managers can't tell if they are from a man or a woman. In a study which anonymized applications from scientists for time on the Hubble Space Telescope, men outperformed women when their gender was known. After anonymization, women outperformed men.[29] We also know that a woman applicant is 30 per cent less likely to be called for a job interview than an identically qualified man.[30]

- We can insist on putting more than one woman on every shortlist. Having only one woman against three men means there is statistically zero chance of her being hired.[31] That is because the ratio is sending the implicit message that a man is more suitable for the job. Adding another woman to the shortlist makes the odds of hiring a woman seventy-nine times greater. There is an even larger effect for people of colour.

- We can make sure there are at least two women on the selection panel. Having only one *decreases* the chance of a woman being hired. That is because the men think that they don't have to worry about diversity; they can delegate it to the woman. And the woman fears that if she champions a female candidate, the men will think that she is being nepotistic.
- We can ask job candidates to perform tasks relevant to the job they are applying for, as well as an interview, and use their performance on the task as a good measure of their suitability.[32]
- We can use structured job interviews in which we ask exactly the same questions in the same order of all candidates and we have a strict marking scheme against the criteria we have already laid down.[33] This helps to reduce unconscious bias, such as hiring men for their potential but women only for their past achievements.
- We can introduce transparency about pay and promotion, which helps employees to be clear about what is involved and incentivizes managers to be objective and evidence-based as they know their decisions will be scrutinized.[34]
- If the employer is big enough, we can appoint diversity managers who are responsible for monitoring recruitment and promotion in the organization and will hold managers to account. These people must be senior and have access to all the data they need.[35]
- We can appoint female 'reverse mentors' to male managers, to help them understand how their behaviour can contribute to the authority gap.

- We can encourage fathers to take shared parental leave and reassure them that it won't affect their promotion chances. Each month a father stays on parental leave increases a mother's earnings.[36] If we pay fathers decently, they are much more likely to take it. After Aviva introduced six months' leave on full pay for both fathers and mothers, new fathers took an average of twenty-one weeks' leave, compared with two weeks the previous year.
- We can recruit 'returners': women who have taken time out of the labour force because of caring responsibilities and are either not employed or working in jobs for which they are overqualified. They are a huge source of talent for employers.
- We can offer mentoring and sponsorship programmes for women and not assume that they are always best served by having another woman looking out for them. It can be helpful for them to have a senior man, who might have more influence over his male colleagues.
- We can set specific targets for progress and ensure that we are tracking our success at meeting them.

What can teachers and places of learning do?

After parents, teachers have the next greatest influence on a child's development. If we want to narrow the authority gap in the next generation, schools and universities have a huge part to play. Girls are doing well academically, and a different book should address why so many white working-class boys are falling behind educationally. But in social and behavioural terms, a lot of the attitudes that entrench the authority gap are formed at school, both via teachers and via peer groups, and reinforced at university. The forces that silence girls, which make them believe they are diligent but not talented, and undermine their confidence and boost boys' entitlement, can be reversed only if teachers, schools and universities become aware of them and do something about them.

So what can teachers and places of learning do?

- Teachers can ask for gender bias training as part of professional development. In a study looking at classes with teachers who had been trained, girls no longer waited to be selected by teachers to speak, and the ratio of boys to girls calling out, which had been very much in the boys' favour, shrank to almost zero.[37]
- Relying on pupils volunteering answers is a recipe for a male-dominated classroom. Girls who know the answer are more likely to wait to be called on, while boys are more apt to shout out. Also, if teachers extend the waiting time after a question by several seconds, more girls, more pupils of colour and more shy children will join the discussion.[38]
- Levelling with the class about this problem can help. Teachers can point out to boys when they are dominating the discussion and demanding more attention.
- Bringing female scientists into the classroom has a powerful impact on girls' enthusiasm for science. And humanizing science, making it relevant to pupils' lives, makes both girls and boys like it more.[39]
- If teachers make a special effort to include girls in computer science, such as having special clubs or times for them to use the computer room, female computer use rockets, sometimes surpassing that of boys.[40] Redecorating computer rooms so that they look less stereotypically male also encourages girls to use them, without discouraging boys.[41]

- Teachers can examine their own biases. One study found seven out of ten male teachers attributing boys' success in technology to talent, while dismissing girls' success as due to luck or diligence.[42] Teachers' bias in favour of boys and against girls in early years of schooling has been shown to hold back girls' careers well into adulthood.[43]
- Teachers can question their ability to judge intelligence. Study after study shows that adults – both teachers and parents – underestimate the intelligence of girls. Teachers also find it harder to identify gifted girls than boys. This may be because gifted girls are more likely to try to hide their ability as they know that boys don't like girls being cleverer than them.[44]
- Teachers can put walls to work. Walls covered with white men send a particular message, which isn't helpful to girls or pupils of colour.
- Schools can examine textbooks and teaching materials for bias. Are science textbooks full of pictures of men in lab coats? If so, girls will feel that science isn't for them.
- Teachers can call out sexism, which is still very much alive in today's schools. In a Girlguiding survey, 81 per cent of female respondents (aged eleven to twenty-one) had witnessed or experienced some sort of sexism in the previous week, mainly from boys their own age.[45] Three out of five had heard jokes or remarks that belittled or degraded women. And 64 per cent of teachers hear sexist language at least once a week.[46] Sexism should be treated as seriously in schools as racism is: it's just as pernicious. Yet in England and Wales, only one in five secondary school teachers have received training in recognizing and tackling sexism as part of their Initial Teacher Education.[47]
- Teachers can crack down on sexual harassment. At mixed-sex schools, 37 per cent of girls have been sexually harassed.[48]

- Teachers can resist separating the class into boys and girls. Every time they do this, they are affirming that boys and girls should be treated differently.
- Teachers can be aware of how and why they discipline children. Are noisy girls punished for being 'disruptive', while boys are let off with 'boys will be boys'?
- Schools can identify a senior teacher as a gender champion who will bring together the whole school in a campaign to challenge gender stereotypes. The governing body should be involved in the campaign, too, to show how seriously the school takes it.
- All subjects should be presented equally to all pupils, with no suggestion that some might be too difficult e.g. physics for girls.
- PHSE lessons should include sessions on gender and diversity, with boys encouraged to understand girls' points of view.
- Schools and teachers should be on the lookout for boys being groomed into extreme misogyny.
- Teachers can work really hard on resisting gender stereotypes. When the BBC conducted an experiment to remove gender stereotypes from seven-year-old schoolchildren's lives for just six weeks, the results were striking. Before the experiment, the girls seriously underestimated their intelligence and had lower self-esteem and confidence than the boys. The boys overestimated their ability and found it hard to express emotions. After six weeks, the self-esteem gap between girls and boys shrank from 8 per cent to 0.2 per cent, the girls' self-motivation rose by 12 per cent and they were 40 per cent more accurate at predicting their test scores. Meanwhile, the boys showed more kindness to others and their bad behaviour fell by 57 per cent.[49]

- University teachers can look at the books and articles they put on their reading lists for students to ensure that they are not overwhelmingly male.
- University teachers can use all the tactics described above for schoolteachers to ensure that female students have an equal say in classes and seminars.
- Universities can make small changes to how they offer STEM subjects and win big rewards in female participation. For instance, the University of California, Berkeley, renamed its course 'The Beauty and Joy of Computing'. They now have 50 per cent women taking the course and the women do as well as the men. Carnegie Mellon University increased the proportion of women taking computer science majors from 4 per cent to 42 per cent in just five years by focusing on real-world applications, no longer requiring high-school experience and allowing students to combine it with other subjects.

What can the media do?

The media have a very special role in helping to narrow the authority gap. For it is what we see all around us in the world that shapes the way we think of women and men. If men are cast as authority figures and women as younger sex objects in TV drama and movies, that pattern will imprint itself on our brains. If the most senior commentators or presenters in newspapers and broadcasting are male, that will cement our assumption that men are more expert than women and carry more authority. If the voices and views of women are given less weight, that sends a powerful message. And if the experts quoted by journalists are usually men, that too suggests that men are more authoritative.

The media also create the climate in which women in public life are judged. If newspapers devote more column inches to a female prime minister's hair or shoes than to her tax policy, her power and authority will be diminished. If their political editors are men (and they usually are), politics and female politicians will always be seen through a male lens.

So what can the media do?

- Broadcasters can allow older, authoritative women to appear on TV. Broadcast regulators, such as Ofcom, should insist on this as part of their diversity policy.
- Newspapers can have more female columnists on serious subjects and more women in senior editing jobs.
- The media can stop using tired old sexist tropes about female politicians and always ask themselves, 'Would I say this about a man?' If not, delete.
- The media can interview as many female experts as men, and keep real-time records of contributors to ensure that this happens.
- Political journalists can ask themselves if they are using double standards and treating female leaders more harshly than men.
- The media can stop commenting so much more on women's appearance than men's. It is very easily done, if there's a will. Editors should always 'flip' the descriptors in articles and ask whether the same would be said of a man. If not, cut it out.
- Advertisers can catch up with real life in their portrayal of women and men.
- The film industry can bring in more female directors and give female characters more agency, dimensions and speaking time.

- Film companies can back films with female leads with as much money, distribution and marketing as films with male leads.

What should governments do?

Many of the obstacles to women achieving equality of respect and authority are due to the way they are treated by other people, whether at home or at work. But there are also structural differences that make it harder for women to rise to equal positions of authority in the world, and most of them are to do with childbearing. In this area, laws and government priorities can definitely help.

Enlightened governments, particularly in Northern Europe, are already doing many of these things. America is still pretty far behind.

So what should governments do?

- Mandate paid maternity leave.
- Offer shared parental leave, at a decent rate of pay, but crucially with the proviso that the father's share is 'use it or lose it': he can't just hand it over to the mother. (Without this proviso, as we have seen in the UK, take-up by fathers is negligible.) If parents of either sex are as likely to take time off when a child is born, employers will be less likely to discriminate against women of childbearing age. And if fathers take leave too, mothers can return more quickly to the workforce.
- Ensure the availability of affordable, quality childcare. The cost of it should be tax-deductible as it is a necessary work expense.
- Give employees the right, wherever possible, to work flexibly.

- Enforce gender quotas where nothing else has worked. Quotas have been successful for getting many more women on to boards. They have been successful in bringing the male/female ratio of Labour MPs to parity. Quotas are not ideal but they are often the least worse option.
- Collect data on time spent on unpaid work outside the workplace and how it is divided between women and men, then use it to inform policy-making and budgeting decisions.
- Force employers to publish data on their gender pay gaps. It does help to narrow them.[50]
- Crack down on extreme misogyny, which is often a gateway to far-right white supremacy.
- Prosecute trolls who threaten violence against women.
- Appoint equal numbers of men and women to the Cabinet. Not only does this ensure that women's voices and concerns are equally heard in a representative democracy; it also normalizes the notion that men and women should have equal authority.

What can the rest of society do?

If we really want to reduce the underlying unconscious bias that makes us more reluctant to accord authority to women than to men, we have to make it more normal and less jarring for women to be in positions of authority. It is already happening, and already – to some extent – feeding through into our unconscious, as we read in Chapter 9. To make it vanish altogether (which may take several generations) we need to deal with the status incongruity problem by making women in leadership no longer incongruous – just as it once seemed very odd to see a woman in a pair of trousers but now it's utterly commonplace and not something we even notice.

Most of the solutions above are aimed at getting more women into leadership positions. This will eventually transform our attitudes towards authoritative women. A real-life law change in India provided a wonderful control experiment for researchers: in 1993, the Indian government passed a constitutional amendment to address the dearth of female leaders in local politics. In each five-year election cycle, one third of villages were randomly selected to appoint a female *pradhan* or chief.

The result? After two cycles of having a female *pradhan* in a village, perceptions of women in leadership improved among both male and female villagers. What is more, parents' aspirations for their daughters increased: they were 45 per cent more likely to want their girls to progress beyond secondary school than parents in villages that had never had a female leader. Meanwhile, the girls themselves had greater ambitions, did better at school and shared the household chores more equally with their brothers.[51]

With each woman in a particular leadership position, it becomes easier for the next, as we become used to the idea. As Jacinda Ardern says, 'None of my doubt arose because of a perception that the New Zealand public wouldn't accept me because I am a woman. That's the difference having two prime ministers like Helen Clark and Jenny Shipley made. I could see that you could be elected and you could be a successful prime minister and be a woman.'[52]

And not just at the prime ministerial level, of course. Once the idea of a woman in any leadership position has been normalized, life becomes a lot easier for her successors. They can be more like their authentic selves. They don't have to pretend to be a man in order to succeed in their organization.

General Sir Nick Carter, the head of Britain's armed forces, sees how difficult this is for women who are still anomalies in a mostly male workforce. 'The tragedy of it,' he told me, 'is that all too often women who succeed feel they have to become men in order to prevail, and that is just really sad. I think we'll know we've been successful when women don't have to act like a man.'[53]

Michelle Bachelet was determined not to let this happen when she became Chile's first female Minister of Defence, overseeing the very military that had tortured her and killed her father under the Pinochet regime. 'When I was new to the ministry of defence,' she told me, 'I was talking to my mom on the telephone, and then the colonel, the chief of staff, came in and asked me something. I said [speaking sweetly], "Oh, please, Mr Colonel, can you do this and this and this?" And my mom said afterwards, "Are you sure they're going to respect you if you speak like that?" And I said, "Look, I have my style. I am like I am. I'm not going to be like a man to be respected. If that's what I need, I don't want that." '[54] It paid off. She subsequently became president.

And we'll also know we've been successful when men feel they have a licence to act a bit more like women. 'I think we should hold men accountable for the same good qualities of leadership that we hold women accountable for,' the American businesswoman Anne Mulcahy told me.[55] 'Things like empathy and sensitivity and being personal and being humble and acknowledging when you're wrong and all those good things that go along with a good leadership profile.'

Not only would this make for better leaders; it would also make for happier men, says the psychotherapist Nick Duffell. 'Men, in general, need to learn to be much more comfortable with their vulnerability. They have to know the paradox that, once you do become comfortable with being vulnerable, you're very powerful actually.'[56]

He warmed to his theme. 'We've got to come together, haven't we? We've got to create something totally different now. So, in a way, we have to do what Jung called "the inner marriage": we have to be doing that inside of us and in society, so men have to get much more in touch with their feminine sides and women come to terms with their masculinity and stop pointing fingers and work together.'

This, I believe, is the nub. Men and women actually work really well together. They complement each other. Their different perspectives synthesize into more interesting ideas and ways of doing things. If it became the default that all levels of all organizations were run jointly by a woman with a male deputy or by a man with a female deputy, we would not only have better, more rounded, more imaginative leadership, but we would also achieve rough gender equality in the process. If we only made mixed-sex leadership the norm, the authority gap would shrink in one generation.

I began this book with a quote from the former Irish President Mary McAleese, and I am going to end with another from her, for she is perhaps the most eloquent woman I have talked to on this important subject. 'If men don't take women equally seriously, we end up with this world that flies on one wing, and I don't know if you've ever seen a bird that tries to fly on one wing?' she asked me. 'It can't get elevation, it can't get direction, it flaps about rather sadly. And that's our world, flapping about rather sadly because of the refusal to use the elevation and the direction and the confidence that comes from flying on two wings.

'And the sad thing is that very often this male wing seems to think it has to spend a lot of effort keeping the other wing down. And that's wasted effort, it's wasted lives. It has caused dysfunction in relationships, it has caused dysfunction in families, in communities, in workplaces, in politics, in international politics, in warfare. That's where we have to understand that when women flourish and their talents and their creativity flourish, then the world flourishes and men flourish.

'We all flourish.'[57]

Acknowledgements

I have two very big thank-yous to make: to All Souls College, Oxford, and to the inspiring – and exceptionally busy – women who allowed me time to interview them about how the authority gap has affected their lives.

It was thanks to All Souls hosting me for a year as a Visiting Fellow that I was able to get my head down and research this book in the most perfect possible surroundings for academic endeavour. The college looks after its Visiting Fellows beautifully, and makes them feel very welcome, but the opportunity to hone my ideas in conversation with some of the brightest minds on the planet was an extraordinary privilege. I would like to thank, in particular, Sir John Vickers, Celia Heyes, Ruth Harris, Sir Keith Thomas, Catriona Seth, Kevin O'Rourke, Cecile Fabre, Amy Singer, Max Harris, Wolfgang Ernst, Sir Noel Malcolm, Anthony Gottlieb, Anthony Geraghty, Lucia Zedner, Dame Angela McLean, Dmitri Levitin, Lisa Lodwick, Clare Bucknell and Tess Little for their encouragement and introductions. I am especially indebted to Edward Mortimer, who encouraged me to apply for the Visiting Fellowship in the first place, and to Lord (William) Waldegrave, who assured me that a book on women's authority was a far more interesting proposition than the boring books on British politics that were my alternative proposals. I am also grateful to Nuffield College, which made me an Associate Member, and Oriel College, which made me a Senior Academic Visitor, allowing me to finish my research and complete the book.

Other academics were very helpful, including Ros Ballaster,

Dorothy Bishop, Sue Dopson, Roger Goodman, Jane Green, Trish Greenhalgh, Gina Neff, Richard Ovenden, Brian Parkinson, Olivia Spiegler and Jon Stokes. I'm grateful to Mahzarin Banaji and Tessa Charlesworth, who shared their research on the IAT with me before it was published. I'd also like to thank Philip Stone of Nielsen Book Research for digging into his database for me, and Ben Page and Kelly Beaver of Ipsos MORI for adding questions on to their surveys.

I have talked to about a hundred women for this book, half of them extremely notable, the other half from a broad range of backgrounds and experience. All have candidly shared their life stories, insights and irritations with me, and I am immensely grateful to them. Not all of these have made it into the book, but they have shaped my thoughts and my writing. I am sorry not to have had the space to do them all justice, but even in a book this long, I had to make some painful choices.

Some of the women I knew already; others were generously introduced to me by friends and acquaintances. I'd particularly like to thank Bonnie Arnold, Cherie Booth, John Dennehy, Giles Edwards, Shari Finkelstein, Bob Flanagan, Princess Sarvath El Hassan, Bea Hollond, Darren Hughes, James Kirkup, Deborah Kolar, Maya Lin, Bobby McDonagh, Justus O'Brien, Andrew Roberts and Karl Rove for putting me in touch with some brilliant women, and Electra Wang for sharing her experiences of young women in China, which sadly didn't make it into the book.

My agent, Will Francis, has been a great source of cheer throughout the writing of this book, and my editor at Transworld, Helena Gonda, put so much meticulous thought into her comments on my first draft that a much better book has, I hope, emerged. Kathryn Lamb drew some wonderful cartoons and Anna Morrison produced the striking cover design. The copy-editing and production team cast expert eyes over my work, and Alison Barrow, Emma

Burton and Ella Horne have been great champions of the book to the outside world.

I am very grateful to Deborah Cameron, the most eminent academic in the field of women and authority, who took time to read the whole manuscript and reassured me that it was neither dodgy nor too simplistic, which is just the sort of imprimatur you want from an academic. Hannah Dawson, Katie Hickman and Norah Perkins were brilliant at raising my morale when I was feeling blue about this project. And Lara von der Brelie was a great help in researching the chapter on intersectionality.

Most of all, though, I'm grateful to my family. My brother William introduced me to some hotshot women in New York and read some draft chapters. My brother Alister was always on the lookout for relevant links to send me. My daughters, Evie and Rosa, listened to me bore on about this book for years, encouraged me, introduced me to their fascinating female friends and, in Evie's case, read and commented on the whole first draft. And my husband Dai, to whom, as an 'unlikely feminist', this book is dedicated, has been a steadfast ally and has supported me for thirty-five years in all the ways I recommend in the last chapter: respecting me equally, valuing my career, sharing the childcare and chores, having genuine pride in my achievements, feeling equally annoyed by every instance of the authority gap he has witnessed, and bringing me a cup of tea and a chocolate biscuit every afternoon during the writing of this book. I am a very lucky woman.

I have probably missed people out, to whom I apologize in advance. I remain grateful to, and inspired by, the many women who have shared their lives and frustrations with me. Any errors, though, are – of course – my own.

Bibliography

Abele, Andree E., and Woiciszke, Bogdan, *Agency and Communion in Social Psychology* (Routledge, 2019).

Abelson, Miriam J., *Men in Place: trans masculinity, race, and sexuality in America* (University of Minnesota Press, 2019).

Adams, Julia; Brückner, Hannah; and Naslund, Cambria, 'Who counts as a notable sociologist on Wikipedia? Gender, race, and the "professor test"', *Socius: Sociological Research for a Dynamic World*, 5 (2019), https://doi.org/10.1177/2378023118823946.

Adams, Renée B.; Kraeussl, Roman; Navone, Marco A.; and Verwijmeren, Patrick, 'Is gender in the eye of the beholder? Identifying cultural attitudes with art auction prices', 6 Dec. 2017, https://ssrn.com/abstract=3083500.

Adegoke, Yomi, and Uviebinené, Elizabeth, *Slay in Your Lane: the black girl bible* (Fourth Estate, 2018).

Aitkenhead, Decca, 'The interview: Everyday Sexism founder Laura Bates on how teenage boys are being raised on a diet of misogyny', *Sunday Times*, 17 Feb. 2019.

Alexander, Anne, 'Why our democracy needs more black political journalists', *Each Other*, 25 Aug. 2020.

Alter, Charlotte, 'Cultural sexism in the world is very real when you've lived on both sides of the coin', *Time*, 2018, https://time.com/transgender-men-sexism/.

Amnesty International, *Troll Patrol Findings*, 2018, https://decoders.amnesty.org/projects/troll-patrol/findings.

Annenberg Inclusion Initiative, *Inequality across 1,300 Popular Films: examining gender and race/ethnicity of leads/co leads from 2007 to 2019*, research brief, 2020, http://assets.uscannenberg.org/docs/aii-inequality-leads-co-leads-20200103.pdf.

Anzia, Sarah F., and Berry, Christopher R., 'The Jackie (and Jill) Robinson effect: why do congresswomen outperform congressmen?', *American Journal of Political Science*, 55: 3 (2011), pp. 478–93.

Artz, Benjamin; Goodall, Amanda H.; and Oswald, Andrew J., 'Do women ask?', *Industrial Relations*, 57: 4 (2018), pp. 611–36.

Ashley, Louise; Duberley, Jo; Sommerlad, Hilary; and Scholarios, Dora, *A Qualitative Evaluation of Non-Educational Barriers to the Elite Professions* (Social Mobility and Child Poverty Commission, 2015).

Audette, Andre P.; Lam, Sean; O'Connor, Haley; and Radcliff, Benjamin, '(E)quality of life: a cross-national analysis of the effect of gender equality on life satisfaction', *Journal of Happiness Studies*, 20 (2019), pp. 2173–88.

Badham, Van, 'A man lost his job for harassing a woman online? Good', *Guardian*, 2 Dec. 2015.

Baird, Julia, 'Women, own your "Dr" titles', *New York Times*, 28 June 2018.

Ballew, Matthew; Marlon, Jennifer; Leiserowitz, Anthony; and Maibach, Edward, *Gender Differences in Public Understanding of Climate Change* (Yale Program on Climate Change Communication, 20 Nov. 2018).

Bamman, David, 'Attention in "By the book"', 27 Aug. 2018, http://people.ischool.berkeley.edu/~dbamman/btb.html.

Barres, Ben A., 'Does gender matter?', *Nature*, 442: 7099 (2006), pp. 133–6.

Barthelemy, Ramon S.; McCormick, Melinda; and Henderson, Charles, 'Gender discrimination in physics and astronomy: graduate student experiences of sexism and gender microaggressions', *Physical Review Physics Education Research*, 12: 2 (2016), pp. 020119-1–14.

Bates, Laura, 'We must act to stop sexism that starts in the classroom', *Independent*, 24 Sept. 2015.

Bauer, Cara C., and Baltes, Boris B., 'Reducing the effect of stereotypes on performance evaluations', *Sex Roles*, 47: 9–10 (2002), pp. 465–76.

Bauer, Nichole M., 'The gendered qualifications gap', *Behavioral Public Policy* blog, 30 July 2020, https://bppblog.com/2020/07/30/the-gendered-qualifications-gap/.

Bauer, Nichole M., *The Gendered Qualifications Gap: why women must be better than men to win political office* (Cambridge University Press, 2020).

Bazelon, Emily, 'A seat at the head of the table', *New York Times*, 21 Feb. 2020.

BBC, 'Churchill tops PM choice', *Newsnight*, 1 Oct. 2008, http://news.bbc.co.uk/1/hi/programmes/newsnight/7647383.stm.

BBC Media Centre, 'No more boys and girls: can our kids go gender free?', 16 Aug. 2017, https://www.bbc.co.uk/mediacentre/proginfo/2017/33/no-more-boys-and-girls.

BBC News, 'Black MP Dawn Butler "mistaken for cleaner" in Westminster', 29 Feb. 2016, https://www.bbc.co.uk/news/uk-england-london-35685169.

BBC Reality Check team, 'Queen bees: do women hinder the progress of other women?', 4 Jan. 2018, https://www.bbc.co.uk/news/uk-41165076.

Beaman, Lori; Duflo, Esther; Pande, Rohini; and Topalova, Petia, 'Female leadership raises aspirations and educational attainment for girls: a policy experiment in India', *Science*, 335: 6068 (2012), pp. 582–6.

Beard, Mary, *Women and Power: a manifesto* (Profile, 2017).

Beattie, Geoffrey W., 'Turn-taking and interruption in political interviews: Margaret Thatcher and Jim Callaghan compared and contrasted', *Semiotica*, 39: 1–2 (1982), pp. 93–114.

Begeny, Christopher T.; Ryan, Michelle K.; Moss-Racusin, Corinne A.; and Ravetz, Gudrun, 'In some professions, women have become well represented, yet gender bias persists—perpetuated by those who think it is not happening', *Science Advances*, 6: 26 (24 June 2020), pp. 1–10.

Beinart, Peter, 'Fear of a female president', *The Atlantic*, Oct. 2016.

Belmi, Peter; Neale, Margaret A.; Reiff, David; and Ulfe, Rosemary, 'The social advantage of miscalibrated individuals: the relationship between social class and overconfidence and its implications for class-based inequality', *Journal of Personality and Social Psychology*, 118: 2 (2019), pp. 254–82.

Benenson, Joyce F.; Markovits, Henry; and Wrangham, Richard, 'Rank influences human sex differences in dyadic cooperation', *Current Biology*, 24: 5 (2014), pp. R190–1.

Bennedsen, Morten; Simintzi, Elena; Tsoutsoura, Margarita; and Wolfenzon, Daniel, *Do Firms Respond to Gender Pay Gap Transparency?*, working paper (National Bureau of Economic Research, 2019).

Bennett, Arnold, *Our Women: chapters on the sex-discord* (Cassell, 1920).

Berkers, P.; Verboord, M.; and Weij, F., ' "These critics (still) don't write enough about women artists": gender inequality in the newspaper coverage of arts and culture in France, Germany, the Netherlands, and the United States, 1955–2005', *Gender and Society*, 30: 3 (2016), pp. 515–39.

Bernard, Philippe; Content, Joanne; Servais, Lara; Wollast, Robin; and Gervais, Sarah, 'An initial test of the cosmetics dehumanization hypothesis: heavy makeup diminishes attributions of humanness-related traits to women', *Sex Roles*, 83: 1 (2020), pp. 315–27.

Berne, Eric, *Games People Play: the psychology of human relationships* (Grove, 1964).

Bhatt, Wasudha, 'The little brown woman: gender discrimination in American medicine', *Gender and Society*, 27: 5 (2013), pp. 659–80.

Bialik, Carl, 'How unconscious sexism could help explain Trump's win', *Five Thirty-Eight*, 21 Jan. 2017.

Bian, Lin; Leslie, Sarah-Jane; and Cimpian, Andrei, 'Evidence of bias against girls and women in contexts that emphasize intellectual ability', *American Psychologist*, 73: 9 (2018), pp. 1139–53.

Bian, Lin; Leslie, Sarah-Jane; and Cimpian, Andrei, 'Gender stereotypes about intellectual ability emerge early and influence children's interests', *Science*, 355: 6323 (2017), pp. 389–91.

Bilton, Isabelle, 'Women are outnumbering men at a record high in universities worldwide', *Study International*, 7 March 2018.

Birger, Jon, 'Xerox turns a new page', CNN *Money Magazine*, 16 March 2004.

Bohnet, Iris; van Geen, Alexandra; and Bazerman, Max, 'When performance trumps gender bias: joint versus separate evaluation', *Management Science*, 62: 5 (2016), pp. 1225–34.

Bosson, Jennifer K., and Vandello, Joseph A., 'Precarious manhood and its links to action and aggression', *Current Directions in Psychological Science*, 20: 2 (2011), pp. 82–6.

Bowles, Hannah Riley; Babcock, Linda; and Lai, Lei, 'Social incentives for gender differences in the propensity to initiate negotiations: sometimes it does hurt to ask', *Organizational Behavior and Human Decision Processes*, 103: 1 (2007), pp. 84–103.

Boyne, John, ' "Women are better writers than men": novelist John Boyne sets the record straight', *Guardian*, 12 Dec. 2017.

Brackett, Marc A.; Rivers, Susan E.; Shiffman, Sara; Lerner, Nicole; and Salovey, Peter, 'Relating emotional abilities to social functioning: a comparison of self-report and performance measures of emotional intelligence', *Journal of Personality and Social Psychology*, 91: 4 (2006), pp. 780–95.

Brazelton, T. Berry, *The Earliest Relationship: parents, infants, and the drama of early attachment* (Da Capo Lifelong, 1991).

Breda, Thomas, and Napp, Clotilde, 'Girls' comparative advantage in reading can largely explain the gender gap in math-related fields', *Proceedings of the National Academy of Sciences of the United States of America*, 116: 31 (2019), pp. 15435–40.

Brescoll, Victoria L., 'Who takes the floor and why: gender, power, and volubility in organizations', *Administrative Science Quarterly*, 56: 4 (2012), pp. 622–41.

Brescoll, Victoria L.; Dawson, Erica; and Uhlmann, Eric Luis, 'Hard won and easily lost: the fragile status of leaders in gender-stereotype-incongruent occupations', *Psychological Science*, 21: 11 (2010), pp. 1640–2.

Breznican, Anthony, '*Little Women* has a little man problem', *Vanity Fair*, 17 Dec. 2019.

Burgess, Adrienne, and Davies, Jeremy, *Cash or Carry? Fathers combining work and care in the UK* (Fatherhood Institute, Dec. 2017).

Burris, Ethan R., 'The risks and rewards of speaking up: managerial responses to employee voice', *Academy of Management Journal*, 55: 4 (2011), pp. 851–75.

Byrnes, James P.; Miller, David C.; and Schafer, William D., 'Gender differences in risk taking: a meta-analysis', *Psychological Bulletin* 125: 3 (1999), pp. 367–83.

Cabrera, M. A., 'Situational judgment tests: A review of practice and constructs assessed', *International Journal of Selection and Assessment*, 9: 1–2 (2001), pp. 103–13.

Cameron, Deborah, 'Imperfect pitch', in *Language: a feminist guide*, 7 June 2019, https://debuk.wordpress.com/2019/06/07/imperfect-pitch/.

Cameron, Deborah, *Language: a feminist guide*, n.d., https://debuk.wordpress.com/.

Cameron, Deborah, 'Tedious tropes: the sexist stereotyping of female politicians', in *Language: a feminist guide*, 18 Dec. 2019, https://debuk.wordpress.com/2019/12/18/tedious-tropes-the-sexist-stereotyping-of-female-politicians/.

Cameron, Deborah, 'Mind the respect gap', 26 Nov. 2017, https://debuk.wordpress.com/2017/11/26/mind-the-respect-gap/.

Cameron, Deborah, and Shaw, Sylvia, *Gender, Power and Political Speech* (Palgrave Macmillan, 2016).

Carli, Linda L., 'Gender differences in interaction style and influence', *Journal of Personality and Social Psychology*, 56: 4 (1989), pp. 565–76.

Carli, Linda L., 'Gender, interpersonal power and social influence', *Journal of Social Issues*, 55: 1 (1999), pp. 81–99.

Carli, Linda L., 'Gender, language and influence', *Journal of Personality and Social Psychology*, 59: 5 (1990), pp. 941–51.

Carli, Linda L.; Lafleur, Suzanne J.; and Loeber, Christopher C., 'Nonverbal behavior, gender, and influence', *Journal of Personality and Social Psychology*, 68: 6 (1995), pp. 1030–41.

Carlson, Daniel L.; Hanson, Sarah; and Fitzroy, Andrea, *The Division of Childcare, Sexual Intimacy, and Relationship Quality in Couples,* working paper (Georgia State University, Sociology Faculty Publications, 2015).

Carmichael, Sarah Green, 'Women at work: make yourself heard', *HBR IdeaCast*, 30 Jan. 2018, https://hbr.org/podcast/2018/01/women-at-work-make-yourself-heard.html.

Carnevale, Anthony P.; Smith, Nicole; and Campbell, Kathryn Peltier, *May the best woman win?* (Georgetown University, 2019).

Carter, Alecia; Croft, Alyssa; Lukas, Dieter; and Sandstrom, Gillian, 'Women's visibility in academic seminars: women ask fewer questions than men', *PLoS One*, 13: 9 (2018), e0202743.

Carter, Jimmy, 'Losing my religion for equality', *The Age*, 15 July 2009.

Carter, Nancy M., and Silva, Christine, *Pipeline's Broken Promise* (Catalyst, 2010).

Casselman, Ben, and Tankersley, Jim, 'Women in economics report rampant sexual assault and bias', *New York Times*, 18 March 2019.

Castilla, Emilio J., 'Accounting for the gap: a firm study manipulating organizational accountability and transparency in pay decisions', *Organization Science*, 26: 2 (2015), pp. 311–33.

Catalyst, *Women and Men in US Corporate Leadership: same workplace, different realities?* (Catalyst, 2004).

Cecco, Leyland, 'Female Nobel Prize winner deemed not important enough for Wikipedia entry', *Guardian*, 3 Oct. 2018.

Chamorro-Premuzic, Tomas, 'Why do so many incompetent men become leaders?', *Harvard Business Review*, Aug. 2013.

Channel 4, 'Winning ad from Channel 4's £1 million Diversity in Advertising award airs tonight', 15 Feb. 2019, https://www.channel4.com/press/news/winning-ad-channel-4s-ps1-million-diversity-advertising-award-airs-tonight.

Charlesworth, T. E. S., and Banaji, M. R., 'Patterns of implicit and explicit attitudes II: long-term change and stability, regardless of group membership', unpublished manuscript (2020).

Charlesworth, T. E. S., and Banaji, M.R., 'Patterns of implicit and explicit stereotypes III: gender-science and gender-career stereotypes reveal long-term change', unpublished manuscript (2020).

Chaudhary, Mayuri, 'New survey reports black women continue to face major barriers to career advancement', *HR Technologist*, 23 Aug. 2019.

Cheng, Joey T.; Tracy, Jessica L.; Ho, Simon; and Henrich, Joseph, 'Listen, follow me: dynamic vocal signals of dominance predict emergent social rank in humans', *Journal of Experimental Psychology General*, 145: 5 (2016), pp. 536–47.

Cheryan, S.; Plaut, V. C.; Davies, P. G.; and Steele, C. M., 'Ambient belonging: how stereotypical cues impact gender participation in computer science', *Journal of Personality and Social Psychology*, 97: 6 (2009), pp. 1045–60.

Cihangir, Sezgin; Barreto, Manuela; and Ellemers, Naomi, 'Men as allies against sexism: the positive effects of a suggestion of sexism by male (vs. female) sources', *SAGE Open*, April–June 2014, https://journals.sagepub.com/doi/pdf/10.1177/2158244014539168.

Cislak, Aleksandra; Formanowicz, Magdalena; and Saguy, Tamar, 'Bias against research on gender bias', *Scientometrics*, 115 (2018), pp. 189–200.

Clift, E., and Brazaitis, T., *Madam President: shattering the last glass ceiling* (Scribner, 2000).

Clinton, Hillary Rodham, *What Happened* (Simon & Schuster, 2016).

Cohan, Peter, 'When it comes to tech start-ups, do women win?', *Forbes*, 25 Feb. 2013.

Cohn, Nate, 'One year from election, Trump trails Biden but leads Warren in battlegrounds', *New York Times*, 4 Nov. 2019.

Colom, Roberto; Juan-Espinosa, Manuel; Abad, Francisco; and García, Luís F., 'Negligible sex differences in general intelligence', *Intelligence*, 28: 1 (2000), pp. 57–68.

Colyard, K. W., 'A breakdown of "By the book" columns shows that male authors are four times more likely to recommend books by men than by women', *Bustle*, 27 Aug. 2018, https://www.bustle.com/p/a-breakdown-of-by-the-book-columns-shows-that-male-authors-are-four-times-more-likely-to-recommend-books-by-men-than-by-women10244493?campaign_id=10&instance_id=10791&segment_id=15163&user_id=7510d1034d46 5a7dab3a390fbd8dc692&regi.

Cook, Nathan J.; Grillos, Tara; and Andersson, Krister P., 'Gender quotas increase the equality and effectiveness of climate policy interventions', *Nature Climate Change*, 9: 4 (2019), pp. 330–4.

Cooke, Rachel, 'Beth Rigby: "I'm going to have to get off telly soon, because I'll be too old" ', *Guardian*, 31 May 2020.

Correll, S., and Simard, C., 'Vague feedback is holding women back', *Harvard Business Review*, April 2016.

Costa, Paul T.; Terracciano, Antonio; and McCrae, Robert R., 'Gender differences in personality traits across cultures: robust and surprising findings', *Journal of Personality and Social Psychology*, 81: 2 (2001), pp. 322–31.

Cowen, Tyler, 'Rebecca Kukla on moving through and responding to the world', 2 Jan. 2019, https://medium.com/conversations-with-tyler/tyler-cowen-rebecca-kukla-feminism-philosophy-efaac99ac2af.

Cowper-Coles, Minna, *Women Political Leaders: the impact of gender on democracy* (Global Institute for Women's Leadership, 2020).

Cox, Daniel, and Jones, Robert P., 'Hillary Clinton opens up a commanding 11-point lead over Donald Trump', 11 Oct. 2016, https://www.prri.org/research/prri-atlantic-oct-11-poll-politics-election-clinton-leads-trump/.

Crespo-Sancho, Catalina, *Can Gender Equality Prevent Violent Conflict?* (World Bank, 28 March 2018).

Criado Perez, Caroline, 'She called the police. They said that there was nothing they could do', *Mamamia*, 13 Nov. 2013, https://www.mamamia.com.au/caroline-criado-perez-cyber-harassment-speech/.

Crockett, Emily, and Frostenson, Sarah, 'Trump interrupted Clinton 51 times at the debate. She interrupted him just 17 times', *Vox*, 27 Sept. 2016, https://www.vox.com/policy-and-politics/2016/9/27/13017666/presidential-debate-trump-clinton-sexism-interruptions.

Croft, Alyssa; Schmader, Toni; Block, Katharina; and Baron, Andrew Scott, 'The second shift reflected in the second generation: do parents' gender roles at home predict children's aspirations?', *Psychological Science*, 25: 7 (2014), pp. 1418–28.

Cross, Emily J., and Overall, Nickola C., 'Women experience more serious relationship problems when male partners endorse hostile sexism', *European Journal of Social Psychology*, 49: 5 (2019), pp. 1022–41.

Crystal, David, and Crystal, Hilary, *Words on Words: quotations about language and languages* (Penguin, 2001).

Cullen, Zoe, and Perez-Truglia, Ricardo, *The Old Boys' Club: schmoozing and the gender gap*, working paper 26530 (National Bureau of Economic Research, 2019).

Cutler, Anne, and Scott, Donia R., 'Speaker sex and perceived apportionment of talk', *Applied Psycholinguistics*, 11: 3 (1990), pp. 253–72.

Damour, Lisa, 'Why girls beat boys at school and lose to them at the office', *New York Times*, 7 Feb. 2019.

Dariel, A.; Kephart, C.; Nikiforakis, N.; and Zenker, C., 'Emirati women do not shy away from competition: evidence from a patriarchal society in transition', *Economic Science Association*, 3: 2 (2017), pp. 121–36.

Darrah, Kim, *A Week in British News: how diverse are the UK's newsrooms?* (Women in Journalism, 2020).

Dean, Steven, 'Understanding gender, disability and the protection gap', *FT Adviser*, 22 July 2019, https://www.ftadviser.com/protection/2019/07/22/understanding-gender-disability-and-the-protection-gap/.

Deedes, W. F., 'Blair's Babes are still on the warpath', *Daily Telegraph*, 14 Aug. 2000.

de Looze, M. E.; Huijts, T.; Stevens, G. W. J. M.; Torsheim, T.; and Vollebergh, W. A. M., 'The happiest kids on Earth: gender equality and adolescent life satisfaction in Europe and North America', *Journal of Youth and Adolescence*, 47 (2018), pp. 1073–85.

del Río, M. F., and Strasser, K., 'Preschool children's beliefs about gender differences in academic skills', *Sex Roles*, 68: 3–4 (2013), pp. 231–8.

Derks, Belle; Ellemers, Naomi; van Laar, Colette; and de Groot, Kim, 'Do sexist organizational cultures create the queen bee?', *British Journal of Social Psychology*, 50: 3 (2011), pp. 519–35.

Derks, Belle; van Laar, Colette; Ellemers, Naomi; and de Groot, Kim, 'Gender-bias primes elicit queen bee responses among senior policewomen', *Psychological Science*, 22: 10 (2011), pp. 1243–9.

Dex, S., and Ward, K., *Parental Care and Employment in Early Childhood* (Equal Opportunities Commission, 2007).

Dezső, Cristian L.; Ross, David Gaddis; and Uribe, Jose, 'Is there an implicit quota on women in top management? A large-sample statistical analysis', *Strategic Management Journal*, 37: 1 (2016), pp. 98–115.

Dixon-Fyle, Sundiatu; Dolan, Kevin; Hunt, Vivian; and Prince, Sara, *Diversity Wins: how inclusion matters* (McKinsey, 19 May 2020), https://www.mckinsey.com/featured-insights/diversity-and-inclusion/diversity-wins-how-inclusion-matters.

Dobbin, F., and Kalev, A., 'Why diversity programs fail', *Harvard Business Review*, July–Aug. 2016, pp. 52–60.

Doran, George H., and Berdahl, J. L., 'The sexual harassment of uppity women', *Journal of Applied Psychology*, 92: 2 (2007), pp. 425–37.

Dunne, G. A., *Lesbian Lifestyles: women's work and the politics of sexuality* (Macmillan, 1997).

Durante, F.; Tablante, C. Bearns; and Fiske, S. T., 'Poor but warm, rich but cold (and competent): social classes in the stereotype content model', *Journal of Social Issues*, 73: 1 (2017), pp. 138–57.

Eagly, Alice H., and Carli, Linda L., 'The female leadership advantage: an evaluation of the evidence', *Leadership Quarterly* 14: 6 (2003), pp. 807–34.

Eagly, Alice, and Carli, Linda L., 'Women and the labyrinth of leadership', *Harvard Business Review*, Sept. 2007.

Eagly, Alice H.; Nater, Christa; Miller, David I.; Kaufmann, Michèle; and Sczesny, Sabine, 'Gender stereotypes have changed: a cross-temporal meta-analysis of US public opinion polls from 1946 to 2018', *American Psychologist*, 18 July 2019.

Eakins, B., and Eakins, G., 'Verbal turn-taking and exchanges in faculty dialogue', in Betty L. Dubois and Isabel Crouch, eds, *Proceedings of the Conference on the Sociology of the Languages of American Women* (Trinity University, 1976).

Eaton, Asia A.; Saunders, Jessica F.; Jacobson, Ryan K.; and West, Keon, 'How gender and race stereotypes impact the advancement of scholars in STEM: professors' biased evaluations of physics and biology post-doctoral candidates', *Sex Roles*, 82: 3–4 (2020), pp. 127–41.

Economist/YouGov, *The Economist/YouGov Poll, October 14–16 2018*, https://d25d2506sfb94s.cloudfront.net/cumulus_uploads/document/7dh1943i0z/econTabReport.pdf.

Eilperin, Juliet, 'White House women want to be in the room where it happens', *Washington Post*, 13 Sept. 2016.

Elborough, Travis, 'Two letters of one's own', *Boundless*, n.d., https://unbound.com/boundless/2019/03/28/virginia-woolf/.

Elizabeth, 'Sex and reading: a look at who's reading whom', *goodreads*, 19 Nov. 2014, https://www.goodreads.com/blog/show/475-sex-and-reading-a-look-at-who-s-reading-whom.

Ellemers, Naomi; Van den Heuvel, Henriette; de Gilder, Dick; Maass, Anne; and Bonvini, Alessandra, 'The underrepresentation of women in science: differential commitment or the queen bee syndrome?', *British Journal of Social Psychology*, 43: 3 (2004), pp. 315–38.

Elliott, Francis, 'Brexit abuse forces MPs to move house', *The Times*, 16 Feb. 2019.

Elliott, James R., and Smith, Ryan A., 'Race, gender and workplace power', *American Sociological Review*, 69: 3 (2004), pp. 365–86.

Ely, Robin J.; Stone, Pamela; and Ammerman, Colleen, 'Rethink what you "know" about high-achieving women', *Harvard Business Review*, Dec. 2014.

Enright, Anne, 'Diary', *London Review of Books*, 21 Sept. 2017.

Eriksson, Mårten; Marschik, Peter B.; Tulviste, Tiia; Almgren, Margareta; Pereira, Miguel Pérez; Wehberg, Sonja; Marjanovič-Umek, Ljubica; Gayraud, Frederique; Kovacevic, Melita; and Gallego, Carlos, 'Differences between girls and boys in emerging language skills: evidence from 10 language communities', *British Journal of Developmental Psychology*, 30: 2 (2012), pp. 326–43.

Esposito, Anita, 'Sex differences in children's conversation', *Language and Speech*, 22: 3 (1979), pp. 213–20.

Esquire, 'The 80 best books every man should read', 1 Apr. 2015.

Evans, Patrick, ' "It's Dr, not Ms," insists historian', BBC News, 15 June 2018, https://www.bbc.co.uk/news/uk-44496876.

Fallon, Amy, 'VS Naipaul finds no woman writer his literary match – not even Jane Austen', *Guardian*, 2 June 2011.

Files, Julia A.; Mayer, Anita P.; Ko, Marcia G.; Friedrich, Patricia; Jenkins, Marjorie; Bryan, Michael J.; Vegunta, Suneela; Wittich, Christopher M.; Lyle, Melissa A.; Melikian, Ryan; Duston, Trevor; Chang, Yu-Hui H.; and Hayes, Sharonne N., 'Speaker introductions at internal medicine grand rounds: forms of address reveal gender bias', *Journal of Women's Health*, 26: 5 (2017), pp. 413–19.

Flood, Alison, 'Readers prefer authors of their own sex, survey finds', *Guardian*, 25 Nov. 2014.

Foran, Clare, 'The curse of Hillary Clinton's ambition', *The Atlantic*, 17 Sept. 2016.

Friskopp, A., and Silverstein, S., *Straight Jobs, Gay Lives* (Touchstone, 1995).

Frith, Bek, 'Women progress when childcare duties are shared more equally', *HR Magazine*, 5 Dec. 2016, https://www.hrmagazine.co.uk/hr-most-influential/profile/women-progress-when-childcare-duties-are-shared-more-equally.

Fulton, Sarah A., 'When gender matters: macro-dynamics and micro-mechanisms', *Political Behaviour*, 36: 3 (2014), pp. 605–30.

Furnham, Adrian; Reeves, Emma; and Budhani, Salima, 'Parents think their sons are brighter than their daughters: sex differences in parental self-estimations and estimations of their children's multiple intelligences', *Journal of Genetic Psychology*, 163: 1 (2002), pp. 24–39.

Gallup, 'State of the American manager: analytics and advice for leaders', 2014, https://www.gallup.com/services/182216/state-american-manager-report.aspx.

Ganley, Colleen M.; George, Casey E.; Cimpian, Joseph R.; and Makowski, Martha B., 'Gender equity in college majors: looking beyond the STEM/non-STEM dichotomy for answers regarding female participation', *American Educational Research Journal*, 55: 3 (2018), pp. 453–87.

Gardiner, Becky; Mansfield, Mahana; Anderson, Ian; Holder, Josh; Louter, Daan; and Ulmanu, Monica, 'The dark side of *Guardian* comments', *Guardian*, 12 April 2016.

Garikipati, Supriya, and Kambhampati, Uma, 'Women leaders are better at fighting the pandemic', *VoxEU/CEPR*, 21 June 2020, https://voxeu.org/article/women-leaders-are-better-fighting-pandemic.

Gaubatz, John A., and Centra, Noreen B., 'Is there gender bias in student evaluations of teaching?', *Journal of Higher Education*, 70: 1 (2000), pp. 17–33.

Gedro, Julie, 'Lesbian presentations and representations of leadership, and the implications for HRD', *Journal of European Industrial Training*, 34: 6 (2010), pp. 552–64.

Gerhart, B., and Rynes, S., 'Determinants and consequences of salary negotiations by male and female MBA graduates', *Journal of Applied Psychology*, 76: 2 (1991), pp. 256–62.

Ghavami, Negin, and Peplau, Letitia Anne, 'An intersectional analysis of gender and ethnic stereotypes: testing three hypotheses', *Psychology of Women Quarterly*, 37: 1 (2012), pp. 113–27.

Gillard, Julia, and Okonjo-Iweala, Ngozi, *Women and Leadership: real lives, real lessons* (Bantam, 2020).

Glass, Ira, 'If you don't have anything nice to say, say it in all caps', *This American Life*, 23 Jan. 2015.

Global Institute for Women's Leadership, 'Women have been marginalised in Covid-19 media coverage', King's College London News Centre, 30 Oct. 2020, https://www.kcl.ac.uk/news/women-have-been-marginalised-in-covid-19-media-coverage.

Global Media Monitoring Project, *Who Makes the News?*, 2015 report, https://whomakesthenews.org/gmmp-2015-reports/.

Gneezy, Uri; Leonard, Kenneth L.; and List, John A., 'Gender differences in competition: evidence from a matrilineal and a patriarchal society', *Econometrica*, 77: 5 (2009), pp. 1637–64.

Gompers, Paul, and Kovvali, Silpa, 'The other diversity dividend', *Harvard Business Review*, July–Aug. 2018.

Good, Jessica; Woodzicka, Julie; and Wingfield, Lylan, 'The effects of gender stereotypic and counter-stereotypic textbook images on science performance', *Journal of Social Psychology*, 150: 2 (2010), pp. 132–47.

Griffeth, Rodger W.; Hom, Peter W.; and Gaertner, Stefan, 'A meta-analysis of antecedents and correlates of employee turnover: update, moderator tests, and research implications for the next millennium', *Journal of Management*, 26: 3 (2000), pp. 463–88.

Griffith, Nicola, 'Books about women don't win big awards: some data', 26 May 2015, https://nicolagriffith.com/2015/05/26/books-about-women-tend-not-to-win-awards/.

Groff, Lauren, 'Lauren Groff: By the book', *New York Times*, 24 May 2018.

Grunspan, Daniel Z.; Eddy, Sarah L.; Brownell, Sara E.; Wiggins, Benjamin L.; Crowe, Alison J.; and Goodreau, Steven M., 'Males underestimate academic performance of their female peers in undergraduate biology classrooms', *PLoS One*, 10 Feb. 2016, https://journals.plos.org/plosone/article?id=10.1371/journal.pone.0148405.

Guinness, Molly, 'Is this the world's sexiest woman (and the most powerful)?', *Guardian*, 17 July 2011.

Gutiérrez y Muhs, Gabriella; Flores Neimann, Yolanda; González, Carmen G.; and Harris, Angela P., *Presumed Incompetent: the intersections of race and class for women in academia* (Utah State University Press, 2012).

Hadjivassiliou, Kari, and Manzoni, Chiara, 'Discrimination and access to employment for female workers with disabilities', 1 June 2017, https://www.researchgate.net/publication/319999703_Discrimination_and_Access_to_Employment_for_Female_Workers_with_Disabilities_DIRECTORATE_GENERAL_FOR_INTERNAL_POLICY_DEPARTMENT_A_ECONOMIC_AND_SCIENTIFIC_POLICY_Study_on_Discrimination_and_Access_to_E.

Handley, Ian M.; Brown, Elizabeth R.; Moss-Racusin, Corinne; and Smith, Jessi L., 'Quality of evidence revealing subtle gender biases in science is in the eye of the beholder', *Proceedings of the National Academy of Sciences of the United States of America*, 112: 43 (2015), pp. 13201–6.

Hannon, John M., and Milkovich, George T., 'The effect of human resource reputation signals on share prices: an event study', *Human Resource Management*, 35: 3 (1996), pp. 405–24.

Harlow, Roxanna, 'Race doesn't matter, but . . . : the effect of race on professors' experiences and emotion management in the undergraduate college classroom', *Social Psychology Quarterly*, 66: 4 (2003), pp. 348–63.

Harvey, Melinda, and Lamond, Julieanne, 'Taking the measure of gender disparity in Australian book reviewing as a field, 1985 and 2013', *Australian Humanities Review*, 60 (2016), pp. 84–107.

Haslanger, Sally, 'Changing the ideology and culture of philosophy: not by reason (alone)', *Hypatia*, 23: 2 (2008), pp. 210–23.

Hazell, Will, 'A-level results: girls tip gender balance in science', *Times Educational Supplement*, 15 Aug. 2019.

Heil, Bill, and Piskorski, Mikolaj, 'New Twitter research: men follow men and nobody tweets', *Harvard Business Review*, June 2009.

Heilman, Madeline E., and Chen, Julie J., 'Same behavior, different consequences: reactions to men's and women's altruistic citizenship behavior', *Journal of Applied Psychology*, 90: 3 (2005), pp. 431–41.

Hekman, David R.; Johnson, Stefanie K.; Foo, Maw-Der; and Yang, Wei, 'Does diversity-valuing behavior result in diminished performance ratings for non-white and female leaders?', *Academy of Management Journal*, 60: 2 (2016), pp. 771–97.

Hengel, Erin, 'Evidence from peer review that women are held to higher standards', *Vox EU/CEPR*, 22 Dec. 2017, https://voxeu.org/article/evidence-peer-review-women are-held-higher-standards.

Hensel, Jana, 'Gender parity in all areas just seems logical', *Die Zeit*, 28 Jan. 2019.

Herbert, Jennifer, and Stipek, Deborah, 'The emergence of gender difference in children's perceptions of their academic competence', *Journal of Applied Developmental Psychology*, 26: 3 (2005), pp. 276–95.

Hess, Amanda, 'Why women aren't welcome on the internet', *Pacific Standard*, 14 June 2017, https://psmag.com/social-justice/women-arent-welcome-internet-72170.

Hewlett, Sylvia Ann, and Green, Tai, *Black Women Ready to Lead* (Centre for Talent Innovation, 2015).

Hockley, Tony, 'Solution aversion', *Behavioral Public Policy* blog, 27 March 2018, https://bppblog.com/2018/03/27/solution-aversion/.

Hodson, Phillip, *Men: An investigation into the emotional male* (Ariel, 1984).

Holmes, Janet, 'Women's talk in public contexts', *Discourse & Society*, 3: 2 (1992), pp. 131–50.

Holter, Øystein Gullvåg, ' "What's in it for men?" Old question, new data', *Men and Masculinities*, 17: 5 (2014), pp. 515–48.

Horowitz, Jason, 'Girding for a fight, McConnell enlists his wife', *New York Times*, 13 May 2014.

Horvath, Michael, and Ryan, Ann Marie, 'Antecedents and potential moderators of the relationship between attitudes and hiring discrimination on the basis of sexual orientation', *Sex Roles*, 48: 3–4, pp. 115–29.

Hosie, Rachel, 'Transgender people reveal how they're treated differently as a man or woman', *Independent*, 13 April 2017.

Howlett, Neil; Pine, Karen J.; Cahill, Natassia; Orakçıoğlu, İsmail; and Fletcher, Ben C., 'Unbuttoned: the interaction between provocativeness of female work attire and occupational status', *Sex Roles*, 72: 3–4 (2015), pp. 105–16.

Huang, Jess; Krivkovich, Alexis; Starikova, Irina; Yee, Lareina; and Zanoschi, Delia, 'Women in the workplace 2019' (McKinsey, Oct. 2019), https://www.mckinsey.com/~/media/McKinsey/Featured%20Insights/Gender%20Equality/Women%20in%20the%20Workplace%202019/Women-in-the-workplace-2019.ashx.

Hughes, Sarah, '*The Golden Rule* by Amanda Craig, review: a perfect murder "mythtery" ', *i*, 3 July 2020.

Hyde, Janet S., and Mertz, Janet E., 'Gender, culture and mathematics performance', *Proceedings of the National Academy of Sciences of the United States of America*, 106: 22 (2009), pp. 8801–7.

Institute of Physics, *It's Different for Girls: the influence of schools* (Institute of Physics, 2012).

Ipsos, Global Institute for Women's Leadership and King's College London, 'International Women's Day 2019: global attitudes towards gender equality', 2019, https://www.kcl.ac.uk/giwl/assets/iwd-giwl-parenting.pdf.

Jacobi, Tonja, and Schweers, Dylan, 'Justice, interrupted: the effect of gender, ideology and seniority at Supreme Court oral arguments', *Virginia Law Review*, 103 (2017), pp. 1379–1485.

Jane, Emma A., *Misogyny Online: a short (and brutish) history* (Sage Swifts, 2016).

Jerrim, John, and Shure, Nikki, 'Young men score highest on "bullshit calculator" ', University College London, 1 April 2019, https://www.ucl.ac.uk/ioe/news/2019/apr/young-men-score-highest-bullshit-calculator.

Johansson, Elly-Ann, *The Effect of Own and Spousal Parental Leave on Earnings*, working paper 2010: 4 (Institute for Labour Market Policy Evaluation, 2010).

Johnson, Stefanie K.; Hekman, David R.; and Chan, Elsa T., 'If there's only one woman in your candidate pool, there's statistically no chance she'll be hired', *Harvard Business Review*, April 2016.

Johnson, Stefanie K., and Kirk, Jessica F., 'Dual-anonymization yields promising results for reducing gender bias: a naturalistic field experiment of applications for Hubble Space Telescope time', *Publications of the Astronomical Society of the Pacific*, 132:034503, March 2020, https://iopscience.iop.org/article/10.1088/1538-3873/ab6ce0/pdf.

Johnson, Wendy; Carothers, Andrew; and Deary, Ian J., 'Sex differences in variability in general intelligence: a new look at the old question', *Perspectives on Psychological Science*, 3: 6 (2008), pp. 518–31.

Jones, Kristen P.; Peddie, Chad I.; Gilrane, Veronica L.; King, Eden B.; and Gray, Alexis L., 'Not so subtle: a meta-analytic investigation of the correlates of subtle and overt discrimination', *Journal of Management*, 42: 6 (2013), pp. 1–26.

Joshi, Aparna; Son, Jooyeon; and Roh, Hyuntak, 'When can women close the gap? A meta-analytic test of sex differences in performance and rewards', *Academy of Management Journal*, 58: 5 (2014), pp. 1516–45.

Jost, John T.; Rudman, Laurie A.; Blair, Irene V.; Carney, Dana R.; Dasgupta, Nilanjana; Glaser, Jack; and Hardin, Curtis D., 'The existence of implicit bias is beyond reasonable doubt: a refutation of ideological and methodological objections and executive summary of ten studies that no manager should ignore', *Research in Organizational Behavior*, 29 (2009), pp. 39–69.

Julé, A., *Gender, Participation and Silence in the Language Classroom: sh-shushing the girls* (Palgrave Macmillan, 2004).

JWT, *The State of Men* (J. Walter Thompson Intelligence, 2013).

Karpf, Anne, *The Human Voice: the story of a remarkable talent* (Bloomsbury, 2011).

Karpowitz, Christopher F., and Mendelberg, Tali, *The Silent Sex: gender, deliberation and institutions* (Princeton University Press, 2014).

Kay, Katty, and Shipman, Claire, *The Confidence Code* (HarperCollins, 2015).

Kerevel, Yann P., and Atkeson, Lonna Rae, 'Reducing stereotypes of female leaders in Mexico', *Political Research Quarterly*, 68: 4 (2015), pp. 732–44.

Killeen, Lauren A.; López-Zafra, Esther; and Eagly, Alice H., 'Envisioning oneself as a leader: comparisons of women and men in Spain and the United States', *Psychology of Women Quarterly*, 30: 3 (2006), pp. 312–22.

Kilmartin, Christopher; Smith, Tempe; Green, Alison; Heinzen, Harriotte; Kuchler, Michael; and Kolar, David, 'A real time social norms intervention to reduce male sexism', *Sex Roles*, 59: 3 (2008), pp. 264–73.

Kimmel, Michael, *Angry White Men: American masculinity at the end of an era* (National, 2013).

Klofstad, Casey A.; Anderson, Rindy C.; and Peters, Susan, 'Sounds like a winner: voice pitch influences perception of leadership capacity in both men and women', *Proceedings of the Royal Society B*, 14 March 2012, https://doi.org/10.1098/rspb.2012.0311.

Knobloch-Westerwick, S.; Glynn, C. J.; and Huge, M., 'The Matilda effect in science communication: an experiment on gender bias in publication quality perceptions and collaboration interest', *Science Communication*, 35: 5 (2013), pp. 603–25.

Knox, Richard, 'Study: men talk just as much as women', NPR, 5 July 2007, https://www.npr.org/templates/story/story.php?storyId=1176218 6&t=159222113880.

Kogan, Deborah Copaken, 'My so-called "post-feminist" life in arts and letters', *The Nation*, 29 April 2013.

Koolen, C. W., *Reading Beyond the Female* (University of Amsterdam, 2018).

Kramer, Andrea S., and Harris, Alton B., 'The persistent myth of female office rivalries', *Harvard Business Review*, Dec. 2019.

Kreager, Alexis, and Follows, Stephen, *Gender Inequality and Screenwriters* (Writers' Union, 2018).

Kristof, Nicholas, 'What the pandemic reveals about the male ego', *New York Times*, 13 June 2020.

Lagarde, Christine, and Ostry, Jonathan D., 'The macroeconomic benefits of gender diversity', *VoxEU/CEPR*, 5 Dec. 2018, https://voxeu.org/article/macroeconomic-benefits-gender-diversity.

Larivière, Vincent; Ni, Chaoqun; Gingras, Yves; Cronin, Blaise; and Sugimoto, Cassidy R., 'Bibliometrics: global gender disparities in science', *Nature*, 504: 7479 (2013), pp. 211–13.

Latu, Ioana M.; Schmid Mast, Marianne; Lammers, Joris; and Bombari, Dario, 'Successful female leaders empower women's behavior in leadership tasks', *Journal of Experimental Social Psychology*, 49: 3 (2013), pp. 444–8.

Lauzen, Martha M., *The Celluloid Ceiling: behind-the-scenes employment of women on the top US films of 2020*, Center for the Study of Women in Television and Film, 2021, https://womenintvfilm.sdsu.edu/wp-content/uploads/2021/01/2020_Celluloid_Ceiling_Report.pdf.

Lauzen, Martha M., 'It's a man's (celluloid) world: portrayals of female characters in the top grossing films of 2019', Center for the Study of Women in Television and Film, 2020, https://womenintvfilm.sdsu.edu/wp-content/uploads/2020/01/2019_Its_a_Mans_Celluloid_World_Report_REV.pdf.

Lauzen, Martha M., 'Thumbs down 2018: film critics and gender, and why it matters', Center for the Study of Women in Television and Film, 2018, https://womenintvfilm.sdsu.edu/wp-content/uploads/2018/07/2018_Thumbs_Down_Report.pdf.

Lavy, Victor, and Sand, Edith, *On the Origins of Gender Human Capital Gaps: short and long term consequences of teachers' stereotypical biases*, working paper (National Bureau of Economic Research, 2015).

Layser, Nikki; Holcomb, Jessie; and Litmann, Justin, 'Twitter makes it worse: political journalists, gendered echo chambers, and the amplification of gender bias', *International Journal of Press/Politics*, 23: 2 (2018), pp. 1–21.

Lean In, *How Outdated Notions about Gender and Leadership are Shaping the 2020 Presidential Race* (Lean In, 2020).

Lean In and McKinsey, *Women in the Workplace 2019* (Lean In and McKinsey, 2019).

Lean In and McKinsey, *Women in the Workplace 2020* (Lean In and McKinsey, 2020).

Leibbrandt, Andreas, and List, John A., 'Do women avoid salary negotiations? Evidence from a large-scale natural field experiment', *Management Science*, 61: 9 (2015), pp. 2016–24.

Levashina, Julie; Hartwell, Christopher J.; Morgeson, Frederick P.; and Campion, Michael A., 'The structured employment interview: narrative and quantitative review of the research literature', *Personnel Psychology*, 67: 1 (2014), pp. 241–93.

Levin, Sam, 'Delta accused of "blatant discrimination" by black doctor after incident on flight', *Guardian*, 13 Oct. 2016.

Levon, Erez, 'Gender, interaction and intonational variation: the discourse functions of high rising terminals in London', *Journal of Sociolinguistics*, 20: 2 (2016), pp. 133–63.

LinkedIn, *Language Matters: how words impact men and women in the workplace*, 2019, https://www.kcl.ac.uk/giwl/assets/linkedin-language-matters-report-final.pdf.

Lipman, Joanne, *Win Win: when business works for women, it works for everyone* (John Murray, 2018).

Livingston, Robert W.; Rosette, Ashleigh Shelby; and Washington, Ella F., 'Can an agentic black woman get ahead? The impact of race and interpersonal dominance on perceptions of female leaders', *Psychological Science*, 23: 4 (2012), pp. 354–58.

Livni, Ephrat, 'Your workplace rewards men more and AI can prove it', *Quartz at Work*, 7 Dec. 2017, https://qz.com/work/1149027/your-workplace-rewards-men-more-and-an-ai-can-prove-it/.

Lopez, German, 'Study: racism and sexism predict support for Trump much more than economic dissatisfaction', *Vox*, 4 Jan. 2017, https://www.vox.com/identities/2017/1/4/14160956/trump-racism-sexism-economy-study.

Loughland, Amelia, 'Female judges, interrupted: a study of interruption behaviour during oral argument in the High Court of Australia', *Melbourne University Law Review*, 43: 2, 2020, pp. 822–51.

Loveday, Leo, 'Pitch, politeness and sexual role: an exploratory investigation into the pitch correlates of English and Japanese politeness formulae', *Language and Speech*, 24: 1 (1981), pp. 71–89.

Lyness, Karen S., and Judiesch, Michael K., 'Are female managers quitters? The relationships of gender, promotions, and family leaves of absence to voluntary turnover', *Journal of Applied Psychology*, 86: 6 (2001), pp. 1167–78.

McBee, Thomas Page, *Amateur: a true story about what makes a man* (Scribner, 2018).

McBee, Thomas Page, 'Until I was a man, I had no idea how good men had it at work', *Quartz*, 13 May 2016, https://qz.com/680275/until-i-was-a-man-i-had-no-idea-how-good-men-had-it-at-work/.

McClean, Elizabeth J.; Martin, Sean R.; Emich, Kyle J.; and Woodruff, Todd, 'The social consequences of voice: an examination of voice type and gender on status and subsequent leader emergence', *Academy of Management Journal*, 61: 5 (2018), pp. 1869–91.

McDaniel, Michael A., and Nguyen, Nhung T., 'Situational judgment tests: a review of practice and constructs assessed', *International Journal of Selection and Assessment*, 9: 1–2 (2001), pp. 103–13.

McDonagh, Margaret, and Fitzsimons, Lorna, *WOMENCOUNT 2020: role, value, and number of female executives in the FTSE 350* (The Pipeline, 2020), https://www.execpipeline.com/wp-content/uploads/2020/07/The-Pipeline-Women-Count-2020-1.pdf.

MacNell, Lillian; Driscoll, Adam; and Hunt, Andrea N., 'What's in a name: exposing gender bias in student ratings of teaching', *Innovative Higher Education*, 40: 4 (2015), pp. 291–303.

Maddocks, Fiona, 'Marin Alsop, conductor of Last Night of the Proms, on sexism in classical music', *Guardian*, 6 Sept. 2013.

Mailer, Norman, *Advertisements for Myself* (Harvard University Press, 1959).

Maliniak, Daniel; Powers, Ryan; and Walter, Barbara F., 'The gender citation gap in International Relations', *International Organization*, 67: 4 (2012), pp. 889–922.

Manne, Kate, *Down Girl: the logic of misogyny* (Oxford University Press, 2017).

Masoud, Tarek; Jamal, Amaney; and Nugent, Elizabeth, 'Using the Qu'rān to empower Arab women? Theory and experimental evidence from Egypt', *Comparative Political Studies*, 49: 12 (2016), pp. 1555–98.

Matschiner, Melannie, and Murnen, Sarah K., 'Hyperfemininity and influence', *Psychology of Women Quarterly*, 23: 3 (1999), pp. 631–42.

Maume, David J.; Hewitt, Belinda; and Ruppanner, Leah, 'Gender equality and restless sleep among partnered Europeans', *Journal of Marriage and Family*, 80: 4 (2018), pp. 1040–58.

Mavin, Sharon, 'Queen bees, wannabees and afraid to bees: no more "best enemies" for women in management?', *British Journal of Management*, 19: S1 (2008), pp. S75–84.

Mavisakalyan, Astghik, and Tarverdi, Yashar, 'Gender and climate change: do female parliamentarians make difference?', *European Journal of Political Economy*, 56: (2019), pp. 151–64.

Mazei, Jens; Hüffmeier, Joachim; Freund, Philipp Alexander; Stuhlmacher, Alice F.; Bilke, Lena; and Hertel, Guido, 'A meta-analysis on gender differences in negotiation outcomes and their moderators', *Psychological Bulletin*, 141: 1 (2015), pp. 85–104.

Meeussen, Loes; van Laar, Colette; and Verbruggen, Marijke, 'Looking for a family man? Norms for men are toppling in heterosexual relationships', *Sex Roles*, 80: 7 (2018), pp. 429–42.

Mehl, Matthias R.; Vazire, Simine; Ramírez-Esparza, Nairán; Slatcher, Richard B.; and Pennebaker, James W., 'Are women really more talkative than men?', *Science*, 317: 5834, 6 July 2007, p. 82.

Merrit, Deborah Jones, 'Bias, the brain, and student evaluations of teaching', *St John's Law Review*, 82: 1 (2008), pp. 251–2.

Miller, David I., and Halpern, Diane F., 'The new science of cognitive sex differences', *Trends in Cognitive Sciences*, 18: 1 (2014), pp. 37–45.

Miller, David I.; Nolla, Kyle M.; Eagly, Alice H.; and Uttal, David H., 'The development of children's gender-science stereotypes: a meta-analysis of 5 decades of US draw-a-scientist studies', *Child Development*, 89: 6 (2018), pp. 1943–55.

Miller, JoAnn, and Chamberlin, Marylin, 'Women are teachers, men are professors: a study of student perceptions', *Teaching Sociology*, 28: 4 (2000), pp. 283–98.

Mills, Eleanor, 'How to deal with men', *British Journalism Review*, 28: 4 (2017), pp. 5–7.

Mills, Eleanor; Hind, Kate; and Quinn, Aine, 'The tycoon and the escort: the business of portraying women in newspapers', *Women in Journalism*, 19 Sept. 2017, https://womeninjournalism.co.uk/the-tycoon-and-the-escort-the-business-of-portraying-women-in-newspapers-2/.

Moran, Caitlin, 'I have 50 facemasks and I intend to use them', *The Times*, 26 June 2020.

Moscatelli, Silvia; Menegatti, Michela; Ellemers, Naomi; Mariani, Marco Giovanni; and Rubini, Monica, 'Men should be competent, women should have it all', *Sex Roles*, 83: 5–6 (2020), pp. 269–88.

Moss-Racusin, Corinne A.; Dovidio, John F.; Brescoll, Victoria L.; Graham, Mark J.; and Handelsman, Jo, 'Science faculty's subtle gender biases favor male students', *Proceedings of the National Academy of Sciences of the United States of America*, 109: 41 (2012), pp. 16474–16479.

Muir, Kate, '*Killing Eve* and the rise of the older screen queen', *Financial Times*, 14 June 2019.

Mulholland, Valentine, 'Why are there disproportionately few female school leaders and why are they paid less than their male colleagues?', *Times Educational Supplement*, 8 March 2018.

Muller-Heyndyk, Rachel, 'Female and younger leaders more susceptible to imposter syndrome', *HR Magazine*, 28 Oct. 2019, https://www.hrmagazine.co.uk/article-details/female-and-younger-leaders-more-susceptible-to-imposter-syndrome.

Murphy, Heather, 'Picture a leader. Is she a woman?', *New York Times*, 16 March 2018.

Murti, Lata, 'Who benefits from the white coat? Gender differences in occupational citizenship among Asian-Indian doctors', *Ethnic and Racial Studies*, 35: 12 (2013), pp. 2035–53.

National Rehabilitation Information Center, 'Working women with disabilities share strategies for countering stereotypes in the workplace', 9 Sept. 2018, https://www.naric.com/?q=en/rif/working-women-disabilities-share-strategies-countering-stereotypes-workplace.

Neimann, Yolanda Flores, *Chicana Leadership: the frontiers reader* (University of Nebraska Press, 2002).

Nelson, Larry R., Jr; Signorella, Margaret L.; and Botti, Karin G., 'Accent, gender, and perceived competence', *Hispanic Journal of Behavioural Sciences*, 38: 2 (2016), pp. 166–85.

Nichols, Catherine, 'Homme de plume: what I learned sending my novel out under a male name', *Jezebel*, 4 Aug. 2015, https://jezebel.com/homme-de-plume-what-i-learned-sending-my-novel-out-und-1720637627.

Nielsen, *African-American Women: our science, her magic*, 21 Sept. 2017, https://www.nielsen.com/us/en/insights/report/2017/african-american-women-our-science-her-magic/#.

Nittrouer, Christine L.; Hebl, Michelle R.; Ashburn-Nardo, Leslie; Trump-Steele, Rachel C. E.; Lane, David M.; and Valian, Virginia, 'Gender disparities in colloquium speakers at top universities', *Proceedings of the*

National Academy of Sciences of the United States of America, 115: 1 (2018), pp. 104–8.

Noland, Marcus, and Moran, Tyler, 'Study: firms with more women in the C-suite are more profitable', *Harvard Business Review*, Feb. 2016.

Nordell, Jessica, 'Why aren't women advancing at work? Ask a transgender person', *New Republic*, 28 Aug. 2014.

Okahana, Hironao, and Zhou, Enyu, *Graduate Enrollment and Degrees: 2007 to 2017* (Council of Graduate Schools, 2018).

O'Kane, Caitlin, ' "Mr Vice President, I'm speaking": Kamala Harris rebukes Pence's interruptions during debate', CBS News, 7 Oct. 2020, https://www.cbsnews.com/news/kamala-harris-mr-vice-president-pence-interruptions/.

Okimoto, Tyler G., and Brescoll, Victoria L., 'The price of power: power-seeking and backlash against female politicians', *Personality and Social Psychology Bulletin*, 36: 7 (2010), pp. 923–36.

Oleszkiewicz, Anna; Pisanski, Katarzyna; Lachowicz-Tabaczek, Kinga; and Sorokowszka, Agnieska, 'Voice-based assessments of trustworthiness, competence, and warmth in blind and sighted adults', *Psychonomic Bulletin and Review*, 24: 3 (2017), pp. 856–62.

Ones, Deniz S., and Viswesvaran, Chockalingam, 'Gender, age and race differences on overt integrity tests: results across four large-scale job applicant data sets', *Journal of Applied Psychology*, 83: 1 (1998), pp. 35–42.

Organisation for Economic Co-operation and Development, *Reading performance (PISA)*, 2019, https://data.oecd.org/pisa/reading-performance-pisa.htm.

Park, G.; Yaden, D. B.; Schwartz, H. A.; Kern, M. L.; Eichstaedt, J. C.; Kosinski, M.; Stillwell, D.; Ungar, L. H.; and Seligman, M. E., 'Women are warmer but no less assertive than men: gender and language on Facebook', *PLoS One*, 11: 5 (2016), e0155885, https:journals.plos.org/plosone/article?id=10.1371%2Fjournal.pone.0155885.

Parke, Ross D., *Fatherhood* (Harvard University Press, 1996).

Parker, Adam, 'Comparative analysis of gender on Twitter in relation to UK politics journalists', *Lissted*, Oct. 2017, https://drive.google.com/file/d/1CoMkc455RvI49Kr0Qzf_xk-dXyovW1A4/view.

Parker, Ceri, ' "When the woman starts talking, the men switch off" – Christine Lagarde on why gender parity is taking so long', World Economic Forum Annual Meeting, 18 Jan. 2017, https:// www.weforum.org/agenda/2017/01/when-the-woman-starts-talking-the-men-switch-off-davos-participants-on-why-gender-parity-is-taking-so-long/.

Parker, Kim, *Women and Leadership: public says women are equally qualified, but barriers persist* (Pew Research Center, 2015).

Paustian-Underdahl, Samantha C.; Walker, Lisa Slattery; and Woehr, David J., 'Gender and perceptions of leadership effectiveness: a meta-analysis of contextual moderators', *Journal of Applied Psychology*, 99: 6 (2014), pp. 1129–45.

Pearce, Edward, 'Sir Gordon Reece', obituary, *Guardian*, 27 Sept. 2001.

Peck, Emily, 'Half the men in the US are uncomfortable with female political leaders', *Huffington Post*, 19 Nov. 2019.

Pemberton, Cecilia; McCormack, Paul; and Russell, Alison, 'Have women's voices lowered across time? A cross sectional study of Australian women's voices', *Journal of Voice*, 12: 2 (1998), pp. 208–13.

Penny, Laurie, 'A woman's opinion is the mini-skirt of the internet', *Independent*, 4 Nov. 2011, https://www.independent.co.uk/voices/commentators/laurie-penny-a-womans-opinion-is-the-mini-skirt-of-the-internet-6256946.html.

Peplau, Letitia Anne, and Fingerhut, Adam, 'The paradox of the lesbian worker', *Journal of Social Issues*, 60: 4 (2004), pp. 719–35.

Perkins, Susan, and Phillips, Katherine W., 'Research: are women better at leading diverse countries than men?', *Harvard Business Review*, Feb. 2019.

Petruzalek, Daniela, 'Gender bias? A transgender perspective!', *Medium*, 17 Jan. 2018, https://medium.com/@danielapetruzalek/gender-bias-a-transgender-perspective-de27f2cd3837.

Petter, Olivia, 'Tackling workplace sexism could boost economy by 35 per cent, IMF chief says', *Independent*, 2 March 2019.

Petts, Richard J.; Knoester, Chris; and Waldfogel, Jane, 'Fathers' paternity leavetaking and children's perceptions of father–child relationships in the United States', *Sex Roles*, 82: 1 (2019), pp. 173–88.

Phillips, Adam, 'Unforgiven', *London Review of Books*, 7 March 2019.

Phillips, Katherine W.; Liljenquist, Katie A.; and Neale, Margaret A., 'Better decisions through diversity', *Kellogg Insight*, 1 Oct. 2010, https://insight.kellogg.northwestern.edu/article/better_decisions_through_diversity.

Pittman, Chavella T., 'Race and gender oppression in the classroom: the experiences of women faculty of color with white male students', *Teaching Sociology*, 38: 3 (2010), pp. 183–96.

Pressner, Kristen, 'Are you biased? I am', TEDxBasel, 30 Aug. 2016, https://www.youtube.com/watch?v=Bq_xYSOZrgU.

Pring, John, 'MP speaks of pride at being dyspraxic at launch of Neurodivergent Labour', 14 Feb. 2019, https://www.disabilitynewsservice.com/mp-speaks-of-pride-at-being-dyspraxic-at launch-of-neurodivergent-labour/.

Pronin, Emily; Lin, Daniel Y.; and Ross, Lee, 'The bias blind spot: perceptions of bias in self versus others', *Personality and Social Psychology Bulletin*, 28: 3 (2002), pp. 369–81.

Propp, Kathleen M., 'An experimental examination of biological sex as a status cue in decision-making groups and its influence on information use', *Small Group Research*, 26: 4 (1995), pp. 451–74.

pwc, *Winning the Fight for Female Talent: how to gain the diversity edge through inclusive recruitment* (pwc, 2017).

Quadlin, Natasha, 'The mark of a woman's record: gender and academic performance in hiring', *American Sociological Review*, 83: 2 (2018), pp. 331–60.

Ramakrishna, Anil; Martínez, Victor R.; Malandrakis, Nikolaos; Singla, Karan; and Narayanan, Shrikanth, 'Linguistic analysis of differences in portrayal of movie characters', *Proceedings of the 55th Annual Meeting of the Association for Computational Linguistics* (Association for Computational Linguistics, 2017), pp. 1669–1678.

Rankin, Sarah, 'New York Times "By the book" interviews', *GitHub*, 14 June 2018, https://github.com/srhrnkn/btb/blob/master/btb.md#new-york-times-by-the-book-interviews.

Rankine, Claudia, *Just Us: an American conversation* (Allen Lane, 2020).

Rattan, Aneeta; Chilazi, Siri; Georgeac, Oriane; and Bohnet, Iris, 'Tackling the underrepresentation of women in media', *Harvard Business Review*, June 2019.

Raw, Louise, 'When women experts are not taken seriously', BBC News, 20 May 2019, https://www.bbc.co.uk/news/uk-48333945.

Reuben, Ernesto; Sapienza, Paola; and Zingales, Luigi, 'How stereotypes impair women's careers in science', *Proceedings of the National Academy of Sciences of the United States of America*, 111: 12 (2014), pp. 4403–8.

Rigby, Jennifer; Newey, Sarah; and Gilbert, Dominic, 'Why do female leaders seem so good at tackling the coronavirus pandemic?', *Daily Telegraph*, 28 April 2020.

Roberts, Laura Morgan; Mayo, Anthony J.; Ely, Robin J.; and Thomas, David A., 'Beating the odds', *Harvard Business Review*, March–April 2018.

Robertson, Katie, 'Kamala Harris cartoon in Murdoch paper is denounced as racist', *New York Times*, 17 Aug. 2020.

Rodionova, Zlata, 'What happened when a man and woman switched names at work for a week', *Independent*, 10 March 2017.

Rosenthal, Cindy Simon, *When Women Lead: integrative leadership in state legislatures* (Oxford University Press, 1998).

Rosette, Ashleigh Shelby; Koval, Christy Zhou; Ma, Anyi; and Livingston, Robert, 'Race matters for women leaders: intersectional effects on agentic deficiencies and penalties', *Leadership Quarterly*, 27: 3 (2016), pp. 429–45.

Rosette, Ashleigh Shelby, and Livingston, Robert W., 'Failure is not an option for black women: effects of organizational performance on leaders with single versus dual-subordinate identities', *Journal of Experimental Social Psychology*, 48: 5 (2012), pp. 1162–7.

Ross, Karen; Boyle, Karen; Carter, Cynthia; and Ging, Debbie, 'Women, men and news: it's life, Jim, but not as we know it', *Journalism Studies*, 19: 6 (2018), pp. 824–45.

Ross, Karen, and Sreberny-Mohammadi, Annabelle, 'Playing house – gender, politics and the news media in Britain', *Media, Culture & Society*, 19: 1 (1997), pp. 101–9.

Rudman, L. A., 'Self-promotion as a risk factor for women: the costs and benefits of counterstereotypical impression management', *Journal of Personality and Social Psychology*, 74: 3 (1998), pp. 629–45.

Sadker, David; Sadker, Myra; and Zittleman, Karen R., *Still Failing at Fairness: how gender bias cheats girls and boys in school and what we can do about it* (Scribner, 2009).

Sadker, Myra, and Sadker, David, *Final Report: project effect (effectiveness and equity in college teaching)* (US Department of Education, 1986).

Sage, Adam, 'How misogyny, infidelity and betrayal destroyed Ségolène Royal's bid to become president of France', *The Times*, 10 Nov. 2018.

SAGE Publications, 'Gay and lesbian job seekers face discrimination', 5 April 2015, https://phys.org/news/2015-04-gay-lesbian-job-seekers-discrimination.html.

Salerno, Jessica M., and Peter-Hagene, Liana C., 'One angry woman: anger expression increases influence for men, but decreases influence for women, during group deliberation', *Law and Human Behaviour*, 39: 6, pp. 581–92.

Sandberg, Sheryl, *Lean In: women, work, and the will to lead* (W. H. Allen, 2013).

Schilt, Kristen, 'Just one of the guys? How transmen make gender visible at work', *Gender and Society*, 20: 4 (2006), pp. 465–90.

Schilt, Kristen, *Just One of the Guys? Transgender men and the persistence of gender inequality* (University of Chicago Press, 2010).

Schumaker, Erin, 'Progressive gender views among teen boys could protect against violence: study', ABC News, 27 Dec. 2019, https://abcnews.go.com/Health/progressive-gender-views-teen-boys-protect-violence-study/story?id=67897133.

Sesko, Amanda K., and Biernat, Monica, 'Prototypes of race and gender: the invisibility of Black women', *Journal of Experimental Social Psychology*, 46: 2 (2010), pp. 356–60.

Shashkevich, Alex, 'Stanford researcher examines how people perceive interruptions in conversation', 2 May 2018, https://news.stanford.edu/press-releases/2018/05/02/exploring-interrion-conversation/.

Shift7, 'Female-led films outperform at box office for 2014–2017', Dec. 2018, https://shift7.com/media-research/.

Siegel, Ed, 'Could a good choice for the BSO turn into a great choice for Boston?', WBUR, 6 Sept. 2019, https://www.wbur.org/artery/2019/09/06/conductor-andris-nelsons-bso-five-years.

Sieghart, Mary Ann, 'Are you taken less seriously than men?', *Mumsnet*, 29 May 2020, https://www.mumsnet.com/Talk/womens_rights/3923344-Are-you-taken-less-seriously-than-men-Contribute-to-my-book.

Sieghart, Mary Ann, 'Why are even women biased against women?', *Analysis*, BBC Radio 4, 28 April 2019, https://www.bbc.co.uk/programmes/articles/312fXcsr5T1V9p509XNMYC4/why-are-even-women-biased-against-women.

Simge, Andı; Selva, Meera; and Nielsen, Rasmus Kleis, 'Women and leadership in the news media 2020: evidence from ten markets', Reuters Institute for the Study of Journalism, 8 March 2020, https://reutersinstitute.politics.ox.ac.uk/sites/default/files/2020-03/Andi_et_al_Women_and_Leadership_in_Media_FINAL.pdf.

Smith, David, 'Women are still a closed book to men', *Guardian*, 29 May 2005.

Smith, Stacy L.; Choueiti, Marc; Yao, Kevin; Clark, Hannah; and Pieper, Katherine, *Inclusion in the Director's Chair: analysis of director gender & race/ethnicity across 1,300 top films from 2007 to 2019* (ReFrame and Annenberg Inclusion Initiative, Jan. 2020).

Smith, Stacy L.; Weber, Rene; Choueiti, Marc; Pieper, Katherine; Case, Ariana; Yao, Kevin; and Lee, Carmen, *The Ticket to Inclusion: gender & race/ethnicity of leads and financial performance across 1,200 popular films* (ReFrame and Annenberg Inclusion Initiative, Feb. 2020).

Snow, Jon, *Maggie & Me*, Channel 4, 8 April 2013, https://www.channel4.com/programmes/maggie-me.

Snyder, Kieran, 'The abrasiveness trap: high-achieving men and women are described differently in reviews', *Fortune*, 26 Aug. 2014.

Snyder, Kieran, 'Boys learn to interrupt. Girls learn to shut up', *Slate*, 14 Aug. 2014, https://slate.com/human-interest/2014/08/child-interruption-study-boys-learn-to-interrupt-girls-as-young-as-4-years-old.html.

Snyder, Kieran, 'How to get ahead as a woman in tech: interrupt men', *Slate*, 23 July 2014, https://slate.com/human-interest/2014/07/study-men-interrupt-women-more-in-tech-workplaces-but-high-ranking-women-learn-to-interrupt.html.

Snyder, Kirk, *The G Quotient: why gay executives are excelling as leaders . . . and what every manager needs to know* (Wiley, 2006).

Soderlind, Laura, 'Lesbians earn more than heterosexual women while gay men lag in wages', 31 March 2015, https://phys.org/pdf347005749.pdf.

Solnit, Rebecca, *Men Explain Things to Me* (Haymarket, 2014).

Sontag, Susan, 'The double standard of aging', *Saturday Review*, 23 Sept. 1972.

Spender, Dale, *Learning to Lose: sexism and education* (Women's Press, 1980).

Steinpreis, Rhea E.; Anders, Katie A.; and Ritzke, Dawn, 'The impact of gender on the review of curricula vitae of job applicants and tenure candidates: a national empirical study', *Sex Roles*, 41: 7–8 (1999), p. 509.

Stephens-Davidowitz, Seth, 'Google, tell me. Is my son a genius?', *New York Times*, 18 Jan. 2014.

Steuter-Martin, Marilla, 'Sue Montgomery calls out gender disparity at city council, one stitch at a time', CBC News, 14 May 2019, https://www.cbc.ca/news/canada/montreal/montreal-city-council-gender-sue-montgomery-1.5135001.

Storage, Daniel; Charlesworth, Tessa; Banaji, Mahzarin; and Cimpian, Andrei, 'Adults and children implicitly associate brilliance with men more than women', *Journal of Experimental Social Psychology*, 90 (2020), art. 104120, https://www.sciencedirect.com/science/article/abs/pii/S0022103120303607.

Storage, Daniel; Horne, Zachary; Cimpian, Andrei; and Leslie, Sarah-Jane, 'The frequency of "brilliant" and "genius" in teaching evaluations predicts the representation of women and African Americans across fields', *PLoS One*, 3 March 2016, https://journals.plos.org/plosone/article?id=10.1371/journal.pone.0150194.

Subtirelu, Nicholas, 'Bashing Hillary Clinton's voice: "screeching", "shrieking", and "shrill" ', *Linguistic Pulse*, 8 Feb. 2016, https://

linguisticpulse.com/2016/02/08/bashing-hillary-clintons-voice-screeching-shrieking-and-shrill/.

Swacker, M., 'The sex of the speaker as a sociolinguistic variable', in Barrie Thorne and Nancy Henley, eds, *Language and Sex: difference and dominance* (Newbury House, 1975).

Tannen, Deborah, 'The truth about how much women talk – and whether men listen', *Time*, 28 June 2017.

Thomas, Sue; Herrick, Rebekah; Franklin, Lori D.; Godwin, Marcia L.; Gnabasik, Eveline; and Schroedel, Jean R., 'Not for the faint of heart: assessing physical violence and psychological abuse against US mayors', *State and Local Government Review*, 51: 1 (2019), pp. 57–67.

Thomas-Hunt, Melissa C., and Phillips, Katherine W., 'When what you know is not enough: expertise and gender dynamics in task groups', *Personality and Social Psychology Bulletin*, 30: 12 (2004), pp. 1585–98.

Tinsley, Catherine H., and Ely, Robin J., 'What most people get wrong about men and women', *Harvard Business Review*, May–June 2018.

Titlow, John Paul, 'These women entrepreneurs created a fake male cofounder to dodge startup sexism', *Fast Company*, 29 Aug. 2017, https://www.fastcompany.com/40456604/these-women-entrepreneurs-created-a-fake-male-cofounder-to-dodge-startup-sexism.

Tivnan, Tom, 'Women dominated the top literary bestsellers last year', *Bookseller*, 15 Jan. 2018.

Tramontana, Mary Katharine, 'Why are men still explaining things to women?', *New York Times*, 9 Sept. 2020.

Travers, Peter, interview with Anne Hathaway, *Popcorn*, 19 April 2017.

Treneman, Ann, 'Media families: 11. the Siegharts', *Independent*, 28 April 1997.

Trix, F., and Psenka, C., 'Exploring the color of glass: letters of recommendation for female and male medical faculty', *Discourse & Society*, 14: 2 (2003), pp. 191–220.

Twenge, Jean M., 'Changes in masculine and feminine traits over time: a meta-analysis', *Sex Roles*, 36: 5–6 (1997), pp. 305–25.

21st Century Fox, Geena Davis Institute on Gender in Media and J. Walter Thompson Intelligence, 'The Scully Effect: I want to believe in STEM', 2020, https://seejane.org/research-informs-empowers/the-scully-effect-i-want-to-believe-in-stem/.

Uhlmann, Eric Luis, and Cohen, Geoffrey L., 'Constructed criteria: redefining merit to justify discrimination', *Psychological Science*, (2005), pubmed.ncbi.nlm.nih.gov/15943674/.

UK Feminista and National Education Union, *'It's Just Everywhere': a study on sexism in schools – and how we tackle it*, 2017, https://ukfeminista.org.uk/wp-content/uploads/2017/12/Report-Its-just-everywhere.pdf.

Universitat Pompeu Fabra, Barcelona, 'Women are 30 percent less likely to be considered for a hiring process than men', 26 March 2019, https://phys.org/news/2019-03-women-percent-hiring-men.html.

University of Sussex, 'Female bosses favour gay and lesbian job-seekers, research finds', 23 Feb. 2017, https://phys.org/news/2017-02-female-bosses-favour-gay-lesbian.html.

Unstereotype Alliance, 'Advertising is out of sync with world's consumers', 2 Oct. 2018, https://www.unstereotypealliance.org/pt/resources/research-and-tools/ipsos-study---advertising-is-out-of-sync-with-worlds-consumers.

van Bezooijen, Reneé, 'Sociocultural aspects of pitch differences between Japanese and Dutch women', *Language and Speech*, 38: 3 (1995), pp. 253–65.

Vedantam, Shankar, *The Hidden Brain* (Spiegel & Grau, 2010).

VIDA, *The 2018 VIDA Count*, 2019, https://www.vidaweb.org/the-count/the-2018-vida-count/.

Voronova, Liudmila, ' "Send pretty girls to the White House": the role of gender in journalists–politicians' interactions', *Journal for Communication Studies*, 7: 2 (2014), pp. 145–72.

Voyer, Daniel, and Voyer, Susan D., 'Gender differences in scholastic achievement: a meta-analysis', *Psychological Bulletin*, 140: 4 (2014), pp. 1174–1204.

Wagner, Claudia; Graells-Garrido, Eduardo; Garcia, David; and Menczer, Filippo, 'Women through the glass ceiling: gender asymmetries in Wikipedia', *EPJ Data Science*, 5: 1 (2016), DOI: 10.1140/epjds/s13688-016-0066-4.

Waterson, Jim, '*Financial Times* tool warns if articles quote too many men', *Guardian*, 14 Nov. 2018.

Wayne, Carly; Valentino, Nicholas; and Oceno, Marzia, 'How sexism drives support for Donald Trump', *Washington Post*, 23 Oct. 2016.

Weinberg, Dana B., and Kapelner, Adam, 'Comparing gender discrimination and inequality in indie and traditional publishing', *PLoS One*, 9 April 2018, https://journals.plos.org/plosone/article?id=10.1371/journal.pone.0195298.

Weiss, Suzannah, 'Is this the only way to escape trolls?', *Bustle*, 7 April 2015, https://www.bustle.com/articles/74778-tweeting-troll-free-is-a-form-of-male-privilege-and-alex-blank-millard-just-proved-it.

Weitz, Rose, 'Women and their hair: seeking power through resistance and accommodation', *Gender and Society*, 15: 5 (2001), pp. 667–86.

West, Candace, 'When the doctor is a "lady": power, status and gender in physician–patient encounters', *Symbolic Interaction*, 7: 1 (1984), pp. 87–106.

Wible, Pamela, 'Her story went viral. But she is not the only black doctor ignored in an airplane emergency', *Washington Post*, 20 Oct. 2016.

Wieckowski, Ania G., 'For women in business, beauty is a liability', *Harvard Business Review*, Nov.–Dec. 2019.

Wikipedia, 'Gender bias on Wikipedia', 2020, https://en.wikipedia.org/wiki/Gender_bias_on_Wikipedia.

Williams, Blair, 'A gendered media analysis of the prime ministerial ascension of Gillard and Turnbull: he's "taken back the reins" and she's "a backstabbing" murderer', *Australian Journal of Political Science*, 52: 4 (2017), pp. 1036–1146.

Williams, Blair E., 'A tale of two women: a comparative gendered media analysis of UK prime ministers Margaret Thatcher and Theresa May', *Parliamentary Affairs*, April 2020, DOI: 10.1093/pa/gsaa008.

Williams, Joan C., 'The 5 biases pushing women out of STEM', *Harvard Business Review*, March 2015.

Williams, Paula Stone, 'I've lived as a man & a woman – here's what I learned', *Youtube*, 17 Dec. 2017, https://www.youtube.com/watch?v=lrYx7HaUlMY&feature=youtu.be.

Willsher, Kim, 'French "boys' club" of journalists accused of bullying women online', *Guardian*, 11 Feb. 2019.

Winkett, Lucy, 'Thank God for women priests!', *The Oldie*, 8 Jan. 2020.

Woetzel, Jonathan; Madgavkar, Anu; Ellingrud, Kweilin; Labaye, Eric; Devillard, Sandrine; Kutcher, Eric; Manyika, James; Dobbs, Richard; and Krishnan, Mekala, *How Advancing Women's Equality Can Add $12 Trillion to Global Growth* (McKinsey, 1 Sept. 2015), https://www.mckinsey.com/featured-insights/employment-and-growth/how-advancing-womens-equality-can-add-12-trillion-to-global-growth.

Woman Interrupted, *Woman Interrupted*, 2020, http://www.womaninterruptedapp.com/en/.

Woolcock, Nicola, 'GCSE results 2019: top grades on the increase in reformed, harder exams', *The Times*, 23 Aug. 2019.

Wright, Oliver, 'Theresa May making £100,000 a speech on lecture circuit', *The Times*, 22 June 2020.

Wu, Alice H., 'Gender stereotyping in academia: evidence from economics job market rumors forum', Aug. 2017, http://

calwomenofecon.weebly.com/uploads/9/6/1/0/96100906/wu_ejmr_
paper.pdf.

Xinyue Xiao, Sonya; Cook, Rachel E.; Martin, Carol Lynn; Nielson,
Matthew G.; and Field, Ryan D.,'Will they listen to me? An examination
of ingroup gender bias in children's communication beliefs', *Sex Roles*,
80: 3–4 (2019), pp. 172–85.

Yang, Jiang; Counts, Scott; Morris, Meredith Ringel; and Hoff, Aaron,
'Microblog credibility perceptions: comparing the United States and
China', ACM Conference on Computer Supported Cooperative Work,
San Antonio, 2013.

Yong, Ed, 'I spent two years trying to fix the gender imbalance in my
stories', *The Atlantic*, 6 Feb. 2018.

Yoon, Carol Kaesuk, 'Scientist at work: Joan Roughgarden; a theorist with
personal experience of the divide between the sexes', *New York Times*, 17
Oct. 2000.

Zimmerman, Don H., and West, Candace, 'Sex roles, interruptions and
silences in conversation', in Barrie Thorne and Nancy Henley, eds,
Language and Sex: difference and dominance (Newbury House, 1975).

Notes

Introduction

1 Interview with the author, 2019.
2 Livni, 'Your workplace rewards men more'.
3 Joshi et al., 'When can women close the gap?'.
4 https://www.lexico.com/definition/authority.
5 Interview with the author, 2019.
6 Interview with the author, 2020.
7 Jost et al., 'The existence of implicit bias'.
8 Interview with the author, 2019.
9 Interview with the author, 2019.
10 Interview with the author, 2019.
11 Interview with the author, 2020.
12 Interview with the author, 2019.
13 Interview with the author, 2019.
14 Interview with the author, 2020.
15 Bennett, *Our Women*.
16 Elborough, 'Two letters of one's own'.
17 Fallon, 'VS Naipaul finds no woman writer his literary match'.
18 Mailer, *Advertisements for Myself*.
19 Ramakrishna et al., 'Linguistic analysis of differences'.
20 Sieghart, 'Why are even women biased against women?'.
21 Furnham et al., 'Parents think their sons are brighter'.
22 Del Río and Strasser, 'Preschool children's beliefs about gender differences'.
23 Bian et al., 'Evidence of bias against girls and women'.
24 Stephens-Davidowitz, 'Google, tell me.'.
25 Furnham et al., 'Parents think their sons are brighter'.
26 Spender, *Learning to Lose*.
27 Manne, *Down Girl*.
28 Interview with the author, 2019.
29 Birger, 'Xerox turns a new page'.
30 Interview with the author, 2019.

31 Lean In and McKinsey, *Women in the Workplace 2019*. A smaller Ipsos-MORI poll also found women experiencing this treatment more than men, but in lower numbers.

Chapter 1

1 Nichols, 'Homme de plume'.
2 Sieghart, 'Why are even women biased against women?'.
3 Nichols, 'Homme de plume'.
4 Sieghart, 'Why are even women biased against women?'.
5 Moss-Racusin et al., 'Science faculty's subtle gender biases'.
6 Sieghart, 'Why are even women biased against women?'.
7 MacNell et al., 'What's in a name'.
8 Steinpreis et al., 'The impact of gender on the review of curricula vitae'.
9 Interview with the author, 2019.
10 Interview with the author, 2019.
11 Interview with the author, 2019.
12 Eriksson et al., 'Differences between girls and boys'.
13 Voyer and Voyer, 'Gender differences in scholastic achievement'.
14 Bilton, 'Women are outnumbering men at a record high'.
15 Okahana and Zhou, *Graduate Enrollment and Degrees*.
16 Colom et al., 'Negligible sex differences in general intelligence'.
17 Johnson et al., 'Sex differences in variability in general intelligence'.
18 Bian et al., 'Gender stereotypes about intellectual ability'.
19 Miller and Halpern, 'The new science of cognitive sex differences'.
20 Hyde and Mertz, 'Gender, culture and mathematics performance'.
21 Ibid.
22 Woolcock, 'GCSE results 2019'.
23 OECD, *Reading performance (PISA)*.
24 Breda and Napp, 'Girls' comparative advantage in reading'.
25 Hazell, 'A-level results'.
26 Institute of Physics, *It's Different for Girls*.
27 Ganley et al., 'Gender equity in college majors'.
28 Barthelemy et al., 'Gender discrimination in physics and astronomy'.
29 Costa et al., 'Gender differences in personality traits across cultures'.
30 Twenge, 'Changes in masculine and feminine traits'.
31 Park et al., 'Women are warmer but no less assertive than men'.
32 Byrnes et al., 'Gender differences in risk taking'.
33 Brackett et al., 'Relating emotional abilities to social functioning'.
34 Ones and Viswesvaran, 'Gender, age and race differences'.

35 Gneezy et al., 'Gender differences in competition'.
36 Catalyst, *Women and Men in US Corporate Leadership*.
37 Killeen et al., 'Envisioning oneself as a leader'.
38 Carter and Silva, *Pipeline's Broken Promise*.
39 Griffeth et al., 'A meta-analysis of antecedents and correlates'.
40 Lyness and Judiesch, 'Are female managers quitters?'.
41 Elliott and Smith, 'Race, gender and workplace power'.
42 Lean In and McKinsey, *Women in the Workplace 2020*.
43 Mulholland, 'Why are there disproportionately few female school leaders'.
44 Ibid.
45 Paustian-Underdahl et al., 'Gender and perceptions of leadership effectiveness'.
46 Eagly and Carli, 'The female leadership advantage'.
47 Rigby et al., 'Why do female leaders seem so good at tackling the coronavirus pandemic?'.
48 Kristof, 'What the pandemic reveals about the male ego'.
49 Garikipati and Kambhampati, 'Women leaders are better at fighting the pandemic'.
50 Interview with the author, 2019.
51 Interview with the author, 2019.
52 Interview with the author, 2019.
53 Interview with the author, 2019.

Chapter 2

1 Vedantam, *The Hidden Brain*.
2 Barres, 'Does gender matter?'.
3 Nordell, 'Why aren't women advancing at work?'.
4 Vedantam, *The Hidden Brain*.
5 Barres, 'Does gender matter?'.
6 Vedantam, *The Hidden Brain*.
7 Interview with the author, 2018.
8 Yoon, 'Scientist at work'.
9 Schilt, *Just One of the Guys?*.
10 Abelson, *Men in Place*.
11 Schilt, 'Just one of the guys?'.
12 Alter, 'Cultural sexism in the world is very real'.
13 Williams, 'I've lived as a man & a woman'.
14 McBee, *Amateur*.
15 Rodionova, 'What happened when a man and woman switched names'.

16 Titlow, 'These women entrepreneurs created a fake male cofounder'.
17 Barres, 'Does gender matter?'.

Chapter 3
 1 Interview with the author, 2020.
 2 Interview with the author, 2020.
 3 Interview with the author, 2019.
 4 Interview with the author, 2020.
 5 Sieghart, 'Why are even women biased against women?'.
 6 Sieghart, 'Why are even women biased against women?'.
 7 Interview with the author, 2019.
 8 Interview with the author, 2019.
 9 Solnit, *Men Explain Things to Me*.
10 Interview with the author, 2019.
11 Interview with the author, 2020.
12 Interview with the author, 2019.
13 Interview with the author, 2020.
14 Sieghart, 'Are you taken less seriously than men?'.
15 Sesko and Biernat, 'Prototypes of race and gender'.
16 Interview with the author, 2020.
17 Interview with the author, 2019.
18 Sieghart, 'Are you taken less seriously than men?'.
19 Interview with the author, 2019.
20 Interview with the author, 2019.
21 Zimmerman and West, 'Sex roles, interruptions and silences'.
22 Jacobi and Schweers, 'Justice, interrupted'.
23 Loughland, 'Female judges, interrupted'.
24 Interview with the author, 2019.
25 West, 'When the doctor is a "lady" '.
26 Manne, *Down Girl*.
27 Shashkevich, 'Stanford researcher examines how people perceive interruptions'.
28 Karpowitz and Mendelberg, *The Silent Sex*.
29 Esposito, 'Sex differences in children's conversation'.
30 Parke, *Fatherhood*.
31 Interview with the author, 2020.
32 Q&A with the author, 2020.
33 Snyder, 'How to get ahead as a woman in tech'.
34 Interview with the author, 2019.

35 Carmichael, 'Women at work'.
36 Interview with the author, 2019.
37 Woman Interrupted, *Woman Interrupted*.
38 Eilperin, 'White House women want to be in the room'.
39 Parker, ' "When the woman starts talking, the men switch off" '.
40 Interview with the author, 2019.
41 Interview with the author, 2020.
42 Interview with the author, 2018.
43 Interview with the author, 2019.
44 Interview with the author, 2019.
45 Interview with the author, 2019.
46 BBC News, 'Black MP Dawn Butler "mistaken for cleaner" '.
47 Interview with the author, 2019.
48 Jones et al., 'Not so subtle'.

Chapter 4

1 Maume et al., 'Gender equality and restless sleep'.
2 Meeussen et al., 'Looking for a family man?'.
3 Hodson, *Men*.
4 Carlson et al., *The Division of Childcare*.
5 Dex and Ward, *Parental Care and Employment*.
6 LinkedIn, *Language Matters*.
7 Burgess and Davies, *Cash or Carry?*.
8 Frith, 'Women progress when childcare duties are shared'.
9 Petts et al., 'Fathers' paternity leavetaking'.
10 Ipsos et al., 'International Women's Day 2019'.
11 JWT, *The State of Men*.
12 Croft et al., 'The second shift reflected in the second generation'.
13 Schumaker, 'Progressive gender views among teen boys'.
14 Interview with the author, 2020.
15 Interview with the author, 2018.
16 Holter, ' "What's in it for men?" '.
17 Gallup, 'State of the American manager'.
18 Interview with the author, 2019.
19 Hengel, 'Evidence from peer review'.
20 Dixon-Fyle et al., *Diversity Wins*.
21 McDonagh and Fitzsimons, *WOMENCOUNT2020*.
22 pwc, *Winning the Fight for Female Talent*.
23 Hengel, 'Evidence from peer review'.

24 Phillips et al., 'Better decisions through diversity'.

25 Cohan, 'When it comes to tech start-ups'.

26 Lagarde and Ostry, 'The macroeconomic benefits of gender diversity'.

27 Woetzel et al., *How Advancing Women's Equality Can Add $12 Trillion*.

28 Anzia and Berry, 'The Jackie (and Jill) Robinson effect'.

29 Crespo-Sancho, *Can Gender Equality Prevent Violent Conflict?*.

30 Audette et al., '(E)quality of life'.

31 De Looze et al., 'The happiest kids on Earth'.

32 Ballew et al., *Gender Differences in Public Understanding of Climate Change*.

33 Cook et al., 'Gender quotas increase the equality and effectiveness'.

34 Mavisakalyan and Tarverdi, 'Gender and climate change'.

Chapter 5

1 Interview with the author, 2019.

2 Muller-Heyndyk, 'Female and younger leaders'.

3 Interview with the author, 2019.

4 Kay and Shipman, *The Confidence Code*.

5 Brazelton, *The Earliest Relationship*.

6 Furnham et al., 'Parents think their sons are brighter'.

7 Ibid.

8 Interview with the author, 2019.

9 Grunspan et al., 'Males under-estimate academic performance'.

10 Bian et al., 'Evidence of bias against girls and women'.

11 Storage et al., 'Adults and children implicitly associate brilliance with men'.

12 Interview with the author, 2019.

13 Casselman and Tankersley, 'Women in economics report rampant sexual assault and bias'.

14 Wu, 'Gender stereotyping in academia'.

15 Haslanger, 'Changing the ideology and culture of philosophy'.

16 Sadker et al., *Still Failing at Fairness*.

17 Julé, *Gender, Participation and Silence*.

18 Damour, 'Why girls beat boys at school'.

19 Cameron, Deborah, 'Mind the respect gap'.

20 Jerrim and Shure, 'Young men score highest on "bullshit calculator"'.

21 Chamorro-Premuzic, 'Why do so many incompetent men become leaders?'.

22 Belmi et al., 'The social advantage of miscalibrated individuals'.

23 Artz, et al., 'Do women ask?'.

24 Bowles et al., 'Social incentives for gender differences'.

25 Gerhart and Rynes, 'Determinants and consequences of salary negotiations'.
26 Bowles et al., 'Social incentives for gender differences'.
27 Rudman, 'Self-promotion as a risk factor for women'.
28 Sandberg, *Lean In*.
29 Cameron, *Language: a feminist guide*.
30 Treneman, 'Media families'.
31 Horowitz, 'Girding for a fight'.
32 Interview with the author, 2019.
33 Interview with the author, 2019.
34 Interview with the author, 2019.
35 Interview with the author, 2019.
36 Interview with the author, 2018.
37 Interview with the author, 2019.
38 Interview with the author, 2018.
39 Interview with the author, 2019.
40 Kay and Shipman, *The Confidence Code*.
41 Interview with the author, 2019.
42 Interview with the author, 2020.
43 Treneman, 'Media families'.

Chapter 6
1 Interview with the author, 2020.
2 Mehl et al., 'Are women really more talkative than men?'.
3 Tannen, 'The truth about how much women talk'.
4 Interview with the author, 2020.
5 Eakins and Eakins, 'Verbal turn-taking and exchanges'.
6 Interview with the author, 2019.
7 Steuter-Martin, 'Sue Montgomery calls out gender disparity'.
8 Carter et al., 'Women's visibility in academic seminars'.
9 Interview with the author, 2020.
10 Latu et al., 'Successful female leaders empower women's behavior'.
11 Interview with the author, 2019.
12 Swacker, 'The sex of the speaker'.
13 Tramontana, 'Why are men still explaining things to women?'.
14 Tinsley and Ely, 'What most people get wrong'.
15 Karpowitz and Mendelberg, *The Silent Sex*.
16 Deedes, 'Blair's Babes are still on the warpath'.
17 Cutler and Scott, 'Speaker sex and perceived apportionment of talk'.
18 Interview with the author, 2019.

19 Brescoll, 'Who takes the floor and why'.
20 Brescoll, 'Who takes the floor and why'.
21 Interview with the author, 2019.
22 Karpf, *The Human Voice*.
23 Cameron, 'Imperfect pitch'.
24 Beard, *Women and Power*.
25 Pemberton et al., 'Have women's voices lowered across time?'.
26 Van Bezooijen, 'Sociocultural aspects of pitch differences'.
27 Loveday, 'Pitch, politeness and sexual role'.
28 Wright, 'Theresa May making £100,000 a speech'.
29 See https://libquotes.com/keith-waterhouse/quote/lby9y5d
30 Interview with the author, 2020.
31 Hensel, 'Gender parity in all areas'.
32 Interview with the author, 2019.
33 Interview with the author, 2020.
34 Oleszkiewicz et al., 'Voice-based assessments of trustworthiness'.
35 Klofstad et al., 'Sounds like a winner'.
36 Cheng et al., 'Listen, follow me'.
37 Interview with the author, 2019.
38 Q&A with the author, 2020.
39 Levon, 'Gender, interaction and intonational variation'.
40 Subtirelu, 'Bashing Hillary Clinton's voice'.
41 Clinton, *What Happened*.

Chapter 7

1 Propp, 'An experimental examination of biological sex as a status cue in decision-making groups'.
2 Thomas-Hunt and Phillips, 'When what you know is not enough'.
3 Burris, 'The risks and rewards of speaking up'.
4 Interview with the author, 2019.
5 Cooke, 'Beth Rigby: "I'm going to have to get off telly soon"'.
6 Carli, 'Gender, interpersonal power and social influence'.
7 Interview with the author, 2019.
8 McClean et al., 'The social consequences of voice'.
9 Livingston et al., 'Can an agentic black woman get ahead?'
10 Matschiner and Murnen, 'Hyperfemininity and influence'.
11 https://pubmed.ncbi.nlm.nih.gov/26322952/
12 Interview with the author, 2020.
13 Cowen, 'Rebecca Kukla on moving through'.

14 Quadlin, 'The mark of a woman's record'.

Chapter 8

1 Layser et al., 'Twitter makes it worse'.
2 Flood, 'Readers prefer authors of their own sex'.
3 Smith, 'Women are still a closed book to men'.
4 Boyne, ' "Women are better writers than men" '.
5 Enright, 'Diary'.
6 Interview with the author, 2019.
7 Interview with the author, 2019.
8 Hughes, '*The Golden Rule* by Amanda Craig'.
9 Interview with the author, 2019.
10 Koolen, *Reading Beyond the Female*.
11 Tivnan, 'Women dominated the top literary bestsellers'.
12 Elizabeth, 'Sex and reading'.
13 Interview with the author, 2020.
14 Griffith, 'Books about women'.
15 Interview with the author, 2020.
16 'The 80 best books every man should read', *Esquire*, 1 April 2015.
17 VIDA, *The 2018 VIDA Count*.
18 Enright, 'Diary'.
19 Interview with the author, 2020.
20 Interview with the author, 2019.
21 Bamman, 'Attention in "By the book" '.
22 Colyard, 'A breakdown of "By the book" columns'.
23 Groff, 'Lauren Groff: By the book'.
24 Weinberg and Kapelner, 'Comparing gender discrimination and inequality'.
25 Interview with the author, 2020.
26 Breznican, '*Little Women* has a little man problem'.
27 Lauzen, 'Thumbs down 2018'.
28 Berkers et al., ' "These critics (still) don't write enough" '.
29 Adams et al., 'Is gender in the eye of the beholder?'.

Chapter 9

1 Travers, interview with Anne Hathaway.
2 Interview with the author, 2020.
3 Sieghart, 'Why are even women biased against women?'.
4 Pressner, 'Are you biased?'.

5 Begeny et al., 'In some professions'.
6 Reuben et al., ''How stereotypes impair women's careers in science'.
7 Charlesworth and Banaji, 'Patterns of implicit and explicit attitudes II'.
8 Email exchange with the author.
9 Interview with the author, 2019.
10 Miller et al., 'The development of children's gender-science stereotypes'.
11 Charlesworth and Banaji, 'Patterns of implicit and explicit stereotypes III'.
12 Good et al., 'The effects of gender stereotypic and counter-stereotypic textbook images'.
13 Interview with the author, 2019.
14 Interview with the author, 2019.
15 Benenson et al., 'Rank influences human sex differences'.
16 Ellemers et al., 'The underrepresentation of women in science'.
17 Derks et al., 'Do sexist organizational cultures create the queen bee?'.
18 Derks et al., 'Gender-bias primes elicit queen bee responses'.
19 BBC Reality Check team, 'Queen bees'.
20 Interview with the author, 2019.
21 Interview with the author, 2019.
22 Kramer and Harris, 'The persistent myth of female office rivalries'.
23 Mavin, 'Queen bees, wannabees and afraid to bees'.
24 Hekman et al., 'Does diversity-valuing behavior result in diminished performance ratings'.
25 Dezső et al., 'Is there an implicit quota'.
26 McDonagh and Fitzsimons, WOMENCOUNT2020.
27 Interview with the author, 2019.
28 Interview with the author, 2019.

Chapter 10
1 Ross et al., 'Women, men and news'.
2 Interview with the author, 2019.
3 Ross et al., 'Women, men and news'.
4 Global Institute for Women's Leadership, 'Women have been marginalised'.
5 Yong, 'I spent two years trying to fix the gender imbalance'.
6 Darrah, A Week in British News.
7 Waterson, 'Financial Times tool warns if articles quote too many men'.
8 Rattan et al., 'Tackling the underrepresentation of women in media'.
9 Global Media Monitoring Project, Who Makes the News?.
10 See https://vtdigger.org/2012/06/17/4th-estate-infographic-womens-voices-arent-heard-in-media-election-coverage/

11 Global Media Monitoring Project, *Who Makes the News?*.
12 Simge et al., 'Women and leadership in the news media 2020'.
13 Mills, 'How to deal with men'.
14 Interview with the author, 2019.
15 Interview with the author, 2019.
16 Willsher, 'French "boys' club" of journalists'.
17 Email to the author, 2019.
18 Alexander, 'Why our democracy needs more black political journalists'.
19 Interview with the author, 2019.
20 Interview with the author, 2019.
21 Cameron and Shaw, *Gender, Power and Political Speech*.
22 Suggested to the author by Evie Prichard.
23 Channel 4, 'Winning ad'.
24 Unstereotype Alliance, 'Advertising is out of sync'.
25 Kreager and Follows, *Gender Inequality and Screenwriters*.
26 Interview with the author, 2020.
27 Ibid.
28 21st Century Fox et al., 'The Scully Effect'.
29 Annenberg Inclusion Initiative, *Inequality across 1,300 Popular Films*.
30 See www.bechdeltest.com.
31 Lauzen, 'It's a man's (celluloid) world'.
32 Lauzen, *The Celluloid Ceiling*.
33 Smith et al., *Inclusion in the Director's Chair*.
34 Shift7, 'Female-led films outperform at box office'.
35 Ibid.
36 Interview with the author, 2019.
37 Carter, 'Losing my religion for equality'.
38 Interview with the author, 2019.
39 Interview with the author, 2019.
40 Winkett, 'Thank God for women priests!'.
41 Interview with the author, 2019.
42 Interview with the author, 2019.
43 Masoud et al., 'Using the Qu'rān to empower Arab women?'.

Chapter 11

1 Interview with the author, 2020.
2 See https://worldwitandwisdom.com/author/gloria-steinem/.
3 Interview with the author, 2019.
4 Carnevale et al., *May the best woman win?*.

5 Sage, 'How misogyny, infidelity and betrayal destroyed'. (Sage, 2018)
6 Interview with the author, 2019.
7 Interview with the author, 2020.
8 Williams, 'A gendered media analysis'.
9 Q&A with the author, 2019.
10 Cohn, 'One year from election'.
11 Cameron, 'Tedious tropes'.
12 Okimoto and Brescoll, 'The price of power'.
13 Wayne et al., 'How sexism drives support for Donald Trump'.
14 Lopez, 'Study'.
15 Q&A with the author, 2019.
16 Snow, *Maggie & Me*.
17 BBC, 'Churchill tops PM choice'.
18 Cameron, 'Tedious tropes'.
19 Williams, 'A tale of two women'.
20 Interview with the author, 2019.
21 Cowper-Coles, *Women Political Leaders*.
22 Perkins and Phillips, 'Research'.
23 Fulton, 'When gender matters'.
24 Kerevel and Atkeson, 'Reducing stereotypes of female leaders'.

Chapter 12
1 Interview with the author, 2019.
2 Interview with the author, 2019.
3 Interview with the author, 2019.
4 Interview with the author, 2020.
5 Interview with the author, 2020.
6 Interview with the author, 2020.
7 Interview with the author, 2020.
8 Interview with the author, 2019.
9 Interview with the author, 2020.
10 Gutiérrez y Muhs et al., *Presumed Incompetent*.
11 Ibid.
12 Ibid.
13 Ibid.
14 Adegoke and Uviebinené, *Slay in Your Lane*.
15 Harlow, 'Race doesn't matter, but . . .'.
16 Miller and Chamberlin, 'Women are teachers, men are professors'.

17 Pittman, 'Race and gender oppression in the classroom'.
18 Interview with the author, 2020.
19 Neimann, *Chicana Leadership*.
20 Hewlett and Green, *Black Women Ready to Lead*.
21 Lean In and McKinsey, *Women in the Workplace 2020*.
22 Nielsen, *African-American Women*.
23 Huang et al., 'Women in the workplace 2019'.
24 Chaudhary, 'New survey reports black women continue to face major barriers'.
25 Williams, 'The 5 biases pushing women out of STEM'.
26 Wible, 'Her story went viral'.
27 Ibid.
28 Murti, 'Who benefits from the white coat?'.
29 Bhatt, 'The little brown woman'.
30 Ghavami and Peplau, 'An intersectional analysis of gender and ethnic stereotypes'.
31 Interview with the author, 2020.
32 Interview with the author, 2020.
33 Williams, 'The 5 biases pushing women out of STEM'.
34 Rosette et al., 'Race matters for women leaders'.
35 Livingston et al., 'Can an agentic black woman get ahead?'.
36 Interview with the author, 2020.
37 Gutiérrez y Muhs et al., *Presumed Incompetent*.
38 Ashley et al., *A Qualitative Evaluation of Non-Educational Barriers*.
39 Nelson et al., 'Accent, gender, and perceived competence'.
40 Durante et al., 'Poor but warm, rich but cold (and competent)'.
41 Belmi et al., 'The social advantage of miscalibrated individuals'.
42 Interview with the author, 2020.
43 Interview with the author, 2018.
44 Soderlind, 'Lesbians earn more than heterosexual women'.
45 Dunne, *Lesbian Lifestyles*.
46 Peplau and Fingerhut, 'The paradox of the lesbian worker'.
47 Friskopp and Silverstein, *Straight Jobs, Gay Lives*.
48 Peplau and Fingerhut, 'The paradox of the lesbian worker'.
49 Snyder, *The G Quotient*.
50 Gedro, 'Lesbian presentations and representations of leadership'.
51 Interview with the author, 2020.
52 Dean, 'Understanding gender'.

53 Lean In and McKinsey, *Women in the Workplace 2020*.
54 National Rehabilitation Information Center, 'Working women with disabilities'.
55 Hadjivassiliou and Manzoni, 'Discrimination and access to employment'.
56 Pring, 'MP speaks of pride at being dyspraxic'.
57 Interview with the author, 2020.
58 Interview with the author, 2020.
59 Interview with the author, 2020.

Chapter 13

 1 Sieghart, 'Why are even women biased against women?'.
 2 Ross and Sreberny-Mohammadi, 'Playing house'.
 3 Guinness, 'Is this the world's sexiest woman'.
 4 Moran, 'I have 50 facemasks and I intend to use them'.
 5 Hensel, 'Gender parity in all areas'.
 6 Clinton, *What Happened*.
 7 Sontag, 'The double standard of aging'.
 8 Interview with the author, 2019.
 9 Beard, *Women and Power*.
10 See https://nationalpost.com/health/why-a-growing-number-of-young-women-are-using-botox-before-they-have-wrinkles
11 Interview with the author, 2019.
12 Interview with the author, 2018.
13 Interview with the author, 2019.
14 Gillard and Okonjo-Iweala, *Women and Leadership*.
15 Interview with the author, 2019.
16 Interview with the author, 2019.
17 Interview with the author, 2020.
18 Interview with the author, 2019.
19 Interview with the author, 2019.
20 Interview with the author, 2019.
21 Rosette et al., 'Race matters for women leaders'.
22 Weitz, 'Women and their hair'.
23 Wieckowski, 'For women in business, beauty is a liability'.
24 Bernard et al., 'An initial test of the cosmetics dehumanization hypothesis'.
25 Howlett et al., 'Unbuttoned'.
26 Interview with the author, 2020.
27 Wieckowski, 'For women in business, beauty is a liability'.
28 Interview with the author, 2019.

29 Interview with the author, 2019.
30 Petter, 'Tackling workplace sexism could boost economy'.
31 Interview with the author, 2019.

Chapter 14

1 Baird, 'Women, own your "Dr" titles'.
2 Evans, ' "It's Dr, not Ms" '.
3 Criado Perez, 'She called the police'.
4 Doran and Berdahl, 'The sexual harassment of uppity women'.
5 Jane, *Misogyny Online*.
6 Weiss, 'Is this the only way to escape trolls?'.
7 Ibid.
8 Penny, 'A woman's opinion'.
9 Gardiner et al., 'The dark side of *Guardian* comments'.
10 Amnesty International, *Troll Patrol Findings*.
11 Thomas et al., 'Not for the faint of heart'.
12 Elliott, 'Brexit abuse forces MPs to move house'.
13 Jane, *Misogyny Online*.
14 Hess, 'Why women aren't welcome on the internet'.
15 Ibid.
16 Badham, 'A man lost his job for harassing a woman online?'.
17 Email to the author, 2019.
18 Bosson and Vandello, 'Precarious manhood'.
19 Phillips, 'Unforgiven'.
20 Interview with the author, 2019.
21 Interview with the author, 2019.
22 Interview with the author, 2020.
23 Kimmel, *Angry White Men*.
24 Interview with the author, 2020.
25 *Economist*/YouGov, *The Economist/YouGov Poll*.

Chapter 15

1 Pronin et al., 'The bias blind spot'.
2 Handley et al., 'Quality of evidence revealing subtle gender biases'.
3 Cislak et al., 'Bias against research on gender bias'.
4 Hockley, 'Solution aversion'.
5 Rankine, *Just Us*.
6 Kilmartin et al., 'A real time social norms intervention'.
7 Ely et al., 'Rethink what you "know" about high-achieving women'.

8 Burgess and Davies, *Cash or Carry?*.
9 Croft et al., 'The second shift reflected in the second generation'.
10 Sadker et al., *Still Failing at Fairness*.
11 Karpowitz and Mendelberg, *The Silent Sex*.
12 Ibid.
13 Ibid.
14 Bauer and Baltes, 'Reducing the effect of stereotypes'.
15 Bohnet et al., 'When performance trumps gender bias'.
16 Heilman and Chen, 'Same behavior, different consequences'.
17 Correll and Simard, 'Vague feedback is holding women back'.
18 Livni, 'Your workplace rewards men more'.
19 Karpowitz and Mendelberg, *The Silent Sex*.
20 Brescoll et al., 'Hard won and easily lost'.
21 Moscatelli et al., 'Men should be competent, women should have it all'.
22 Uhlmann and Cohen, 'Constructed criteria'.
23 Trix and Psenka, 'Exploring the color of glass'.
24 Bazelon, 'A seat at the head of the table'.
25 Cullen and Perez-Truglia, *The Old Boys' Club*.
26 Cihangir et al., 'Men as allies against sexism'.
27 Eagly and Carli, 'Women and the labyrinth of leadership'.
28 Leibbrandt and List, 'Do women avoid salary negotiations?'.
29 Johnson and Kirk, 'Dual-anonymization yields promising results'.
30 Universitat Pompeu Fabra, Barcelona, 'Women are 30 percent less likely to be considered'.
31 Johnson et al., 'If there's only one woman in your candidate pool'.
32 Cabrera, 'Situational judgment tests'.
33 Levashina et al., 'The structured employment interview'.
34 Castilla, 'Accounting for the gap'.
35 Dobbin and Kalev, 'Why diversity programs fail'.
36 Johansson, *The Effect of Own and Spousal Parental Leave on Earnings*.
37 Sadker and Sadker, *Final Report*.
38 Sadker et al., *Still Failing at Fairness*.
39 Ibid.
40 Ibid.
41 Cheryan et al., 'Ambient belonging'.
42 Ibid.
43 Lavy and Sand, *On the Origins of Gender Human Capital Gaps*.
44 Ibid.
45 Bates, 'We must act to stop sexism'.

46 UK Feminista and National Education Union, *'It's Just Everywhere'*.

47 Ibid.

48 Ibid.

49 BBC Media Centre, 'No more boys and girls'.

50 Bennedsen et al., *Do Firms Respond to Gender Pay Gap Transparency?*.

51 Beaman et al., 'Female leadership raises aspirations'.

52 Gillard and Okonjo-Iweala, *Women and Leadership*.

53 Interview with the author, 2020.

54 Interview with the author, 2019.

55 Interview with the author, 2019.

56 Interview with the author, 2019.

57 Interview with the author, 2019.

Index

Index

Index

Index

Index

Index

ABOUT THE AUTHOR

Mary Ann Sieghart spent twenty years as Assistant Editor and columnist at *The Times* and won a large following for her columns on politics, economics, feminism, parenthood and life in general. She has presented many programmes on BBC Radio 4, such as *Start the Week, Profile, Analysis* and *One to One*. She chaired the revival of *The Brains Trust* on BBC2 and recently spent a year as a Visiting Fellow of All Souls College, Oxford. She has chaired the Social Market Foundation think tank and sits on numerous boards.